Prai[illegible]
Littl[illegible]

"Perceptive . . . a particularly transp[illegible] Chu vividly sketches the differences between the education structures of China and the United States in terms that will make readers ponder what they actually think about rote memorization, and parents question their preferences for their own children." —*New York Times*, Editors' Choice

"No reporter has gone as deep into what makes Chinese and American schools different today, or given more reasons we should not copy the Chinese. Yet her rollicking account has hope for both cultures because they share a deep interest in what children learn." —*Washington Post*

"Undoubtedly revealing, fascinating, and filled with aha moments." —*Christian Science Monitor*

"Excellent and absolutely fascinating." —*Business Insider*

"Five Stars. A rollicking read . . . intimate, informative, funny and, yes, at times shocking." —*South China Morning Post*

"A complex view about both what's wrong with the Chinese system and what the American system might learn from it." —*USA Today*

"A perceptive and personal story. . . . A well-balanced look at the school system and at Chinese society as a whole." —*Minneapolis Star Tribune*

"A powerful achievement." —*Los Angeles Review of Books*

"Written with the grace of an elegant novel and the precision of a gripping, nonfiction history book." —Heartland Institute

"Mixing personal anecdotes, observations of Chinese classrooms, interviews with parents and students, and thought-provoking facts about Chinese education, the author reveals how *yingshi jiaoyu*—high-stakes testing—has created a culture of stress and conformity. . . . Chu lets readers consider what skills a twenty-first-century student needs and offers insight on the future of global education."
—*BookPage*

"An investigative look at the Chinese educational system and how it produces such a large number of high-performing students."
—*Book Riot*

"This provocative investigation examines cultural differences between the East and West, and the benefits and shortcomings of how both approach education."
—*Real Simple,* Best of the Month

"[Chu] writes with verve about the hypercompetitive and conformist education that led her son to praise Chairman Mao in song and endure being force-fed food he disliked, and yet she acknowledges the little boy's growing confidence and ingenuity . . . and ultimately grapple[s] with the universal parental question of what it means to raise a successful child."
—*National Book Review*

"This book had me at page one! Whip smart, hilariously funny, and shocking. A must-read."
—Amy Chua, author of *The Battle Hymn of the Tiger Mother* and *The Triple Package*

"So compelling I read it in three days flat."
—Michelle Rhee, founder of StudentsFirst and former chancellor of Washington, DC schools

"*Little Soldiers* is the best book I've read about education in China. . . . [Chu] tells this personal story with great insight and humor, and it's combined with first-rate research into the current state of education in China."
—Peter Hessler, staff writer at *The New Yorker* and author of *River Town*, *Oracle Bones*, and *Country Driving*

"*Little Soldiers* is a book that will endure. With honesty and a terrific sense of humor, Lenora Chu has produced not only an intimate portrait of raising a family far from home, but also the most lucid and grounded account of modern Chinese education that I've ever seen. She brilliantly tests our notions of success and creativity, grit and talent, and never shrinks from her conclusions."

—Evan Osnos, staff writer at *The New Yorker* and National Book Award–winning author of *Age of Ambition*

"This is a rare look inside the gates of Chinese schools that helps demystify many traits and behaviors of the Chinese people."

—Deborah Fallows, contributing writer for *The Atlantic* and author of *Dreaming in Chinese*

"As Lenora Chu takes us along on her own adventure in parenting, she affords us, not only an insider's view of China, but an exploration of people- and society-making at its most foundational. Riveting, provocative and unflinchingly candid, *Little Soldiers* is a must-read for parents, educators, and global citizens alike."

—Gish Jen, author of *The Girl at the Baggage Claim*

"I couldn't put this book down. It's a game changer that challenges our tendency to see education practices in black and white."

—Madeline Levine, PhD, author of *New York Times* bestsellers *The Prince of Privilege* and *Teach Your Children Well*

"Chu's fascinating storytelling urges the reader to ask questions like, 'Do the ends justify the means?' 'Is a child's life for a parent or government to dictate, or is it their own?' These questions and more lie at the heart of Chu's important book, which is necessary reading for educators, parents, and anyone interested in shaping the character and capabilities of the next generation of Americans."

—Julie Lythcott-Haims, former Stanford dean of freshmen, and *New York Times* bestselling author of *How to Raise an Adult*

"Gripping, perceptive, honest, revealing, but—above all—deeply thoughtful."

—Marc Tucker, National Center on Education and the Economy

"Lenora Chu, a gifted journalist, has written a fascinating comparison of the US and Shanghai education systems. *Little Soldiers* offers important insights into the strengths and weaknesses of each. There is much to be learned here about the elements of a better education system for the twenty-first century."

—Tony Wagner, expert in residence, Harvard University Innovation Lab and author of *The Global Achievement Gap and Creating Innovators*

"Chu's narrative is told with the honesty of a journalist, allowing readers to understand the conclusions she draws from her journey but also to form their own view of Chinese education. For anyone who wishes to expand their understanding about Chinese society and its impact on education."

—*Library Journal* (starred review), Editors' Fall Pick

"Chu opens a window on to the complex world of communist China and its competitive methodology, which helps raise highly efficient, obedient, intelligent children, but also squelches individualism and spontaneous creativity from the beginning. . . . An informative, personal view of the Chinese and their educational system that will have many American readers cringing at the techniques used by the Chinese to create perfect students." —*Kirkus Reviews*

"The lively anecdotes, scenes, and conversations that Chu relates while describing her encounters with the Chinese education system will amuse or appall Western readers. . . . By the end, the successes of Chu's son . . . persuade her that, going forward, the global ideal is a blend of Chinese rigor and Western individuality, whatever that might look like." —*Publishers Weekly*

"This engaging narrative is personalized by Chu's often humorous recollections of attending American schools as the daughter of immigrants. *Little Soldiers* offers fascinating peeks inside the world's largest educational system and at the future intellectual "soldiers" American kids will be facing." —*Booklist*

LITTLE SOLDIERS

LITTLE SOLDIERS

AN AMERICAN BOY, A CHINESE SCHOOL, AND THE GLOBAL RACE TO ACHIEVE

LENORA CHU

HARPER

NEW YORK • LONDON • TORONTO • SYDNEY

HARPER

A hardcover edition of this book was published in 2017 by HarperCollins Publishers.

FIRST HARPER PAPERBACKS EDITION PUBLISHED 2018.

Designed by Fritz Metsch
Map by James Sinclair

Library of Congress Cataloging-in-Publication Data has been applied for.

ISBN 978-0-06-236786-0 (pbk.)

20 21 22 LSC 10 9 8 7 6 5 4 3 2

For my parents

I am a little soldier, I practice every day.
I raise my binoculars, I see things clearly.
I take a wooden gun—bang, bang, bang!
I drive a small gunboat—boom boom boom!
I ride as a cavalryman—go go go!
I am a little soldier, I practice every day.
One-two-one, one-two-one, Let us forward march!
FOR . . . WARD . . . MARCH!

—A song taught in Chinese kindergarten

CONTENTS

PART III: CHINESE LESSONS

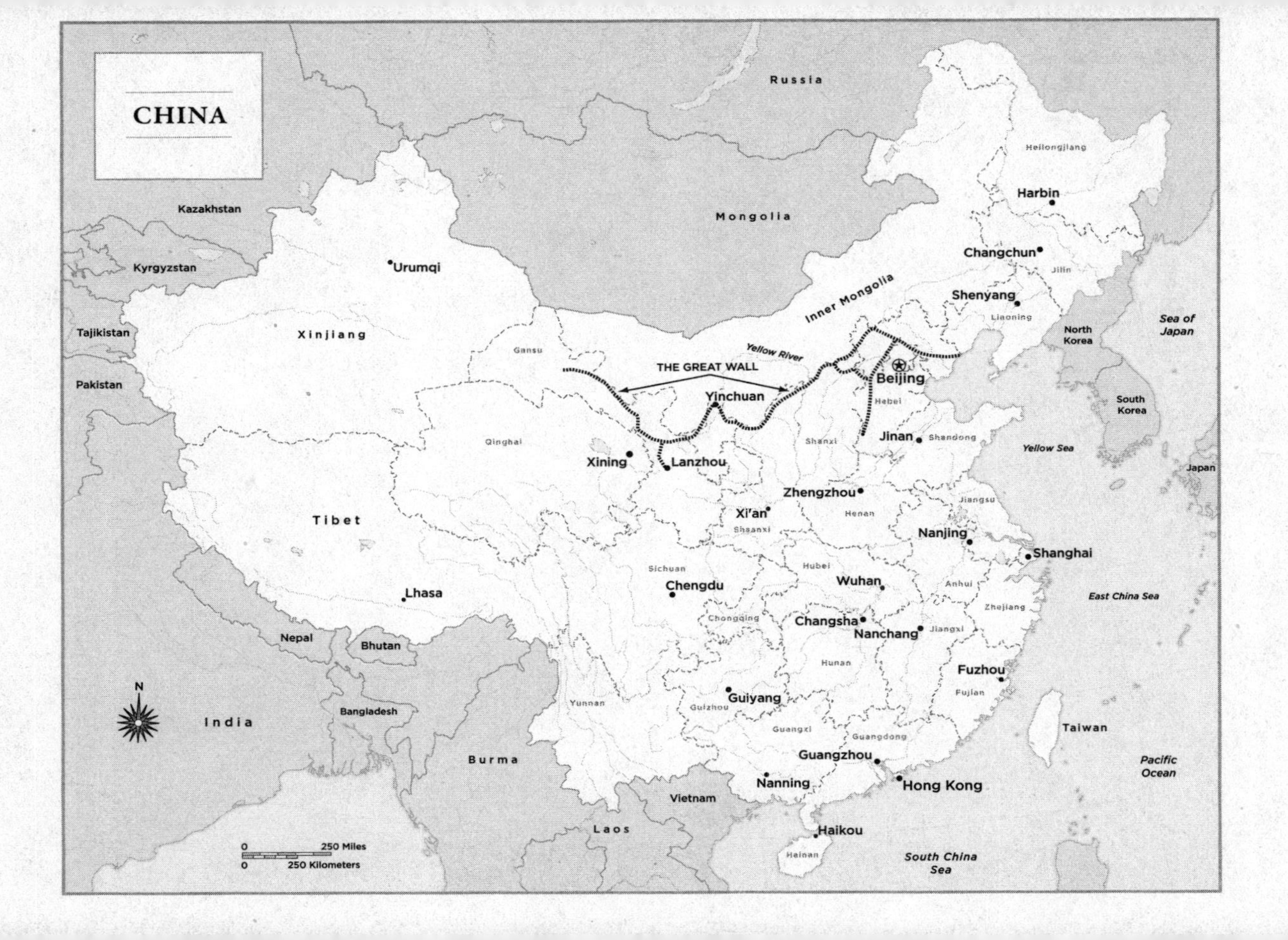
CHINA
Russia
Mongolia
Kazakhstan
Kyrgyzstan
Tajikistan
Pakistan
Urumqi
Xinjiang
Tibet
Lhasa
Nepal
Bhutan
Bangladesh
India
Burma
Laos
Vietnam
N
0 250 Miles
0 250 Kilometers
Gansu
Qinghai
THE GREAT WALL
Yellow River
Inner Mongolia
Yinchuan
Xining
Lanzhou
Heilongjiang
Harbin
Changchun
Jilin
Shenyang
Liaoning
North Korea
South Korea
Sea of Japan
Japan
Yellow Sea
Beijing
Hebei
Jinan
Shandong
Shanxi
Zhengzhou
Henan
Xi'an
Shaanxi
Jiangsu
Nanjing
Shanghai
East China Sea
Sichuan
Chengdu
Hubei
Wuhan
Anhui
Zhejiang
Chongqing
Changsha
Nanchang
Jiangxi
Hunan
Fuzhou
Fujian
Taiwan
Pacific Ocean
Yunnan
Guizhou
Guiyang
Guangxi
Guangdong
Guangzhou
Nanning
Hong Kong
Haikou
Hainan
South China Sea

LITTLE SOLDIERS

PROLOGUE

THE RED STAR

It seemed like a good idea at the time. When my little boy was three years old, I enrolled him in a state-run public school in Shanghai, China's largest city of twenty-six million people.

We're Americans living and working in China, and the Chinese school system is celebrated for producing some of the world's top academic achievers. We were harried, working parents—I'm a writer, my husband's a broadcast journalist—motivated by thoughts of "When in Rome. . ." as well as a desire for a little Chinese-style discipline for our progeny. Our son would also learn Mandarin, the most widely spoken language in the world. Excuse me for thinking, What was not to like?

It seemed an easy decision. And two blocks from our home in downtown Shanghai was Soong Qing Ling, *the* school, as far as posh Chinese urbanites were concerned. Soong Qing Ling Kindergarten educated the three- to six-year-old children of ranking Communist Party officials, wealthy entrepreneurs, real estate magnates, and celebrities. Strolling past on weekends, I sometimes spotted young Chinese parents gazing through the school's entrance gates, as if daydreaming about a limitless future for their child. Inside a culture where the early years are considered so critical that there's a common saying: 不要输在起跑线上, or "Don't lose at the starting gate," we figured Soong Qing Ling was among the best education experiences China had to offer.

A transformation began almost immediately. After the school year began, I noticed my normally rambunctious toddler developing into a proper little pupil. Rainey faithfully greeted his teacher with "*Laoshi zao*!"—Good morning, Teacher! He began heeding the teacher's every command, patiently waiting his turn in lines, and performing little duties around the house when asked.

He also began picking up Mandarin, along with a dose of what the culture valued: hard work and academics. One day, Rainey tried to decipher the meaning of a Chinese phrase he'd heard in the classroom.

"What does *congming* mean?" he asked me, large brown eyes wide.

"*Congming* means 'smart,' Rainey," I responded.

"Oh, good. I want to be smart," he said, bobbing his head. He wrinkled his nose. "What does *ke ai* mean?"

"That means 'cute,'" I said, and his eyes widened.

"Oh! I don't want to be cute. I want to be smart," Rainey said.

One afternoon, he emerged from school with a shiny red star plastered to his forehead.

"Who gave you that star?" I asked my son.

"My teacher! I was good in school," Rainey chirped, glancing up at me as I studied his face.

"What do you get a red star for?" I ventured, always curious about his classroom environment. "Do you get it if you run fast?"

Rainey laughed, a hearty guffaw that emanated from his tiny belly and up through his throat, as if I'd just uttered the most ridiculous thing he'd ever heard.

"Mom, I *never* get a star for running in the classroom," he said, smirking, large, brown eyes dancing. "I get it for sitting still."

Sitting still? I immediately recognized the error in my assumptions.

In America, a student might be rewarded for extraordinary effort or performance, for rising a head above the rest. In China, you get a star for blending in and for doing as you're told. It was America's celebrity culture versus China's model citizen; standing out versus fitting in; individual excellence versus the merit of collective behavior.

The Chinese way was certainly familiar to me, the daughter of Chinese immigrants to America. Yet I'm also a product of US public schools and their culture of personal choice, and as a parent, I wanted good habits instilled for the right reasons, and delivered with a firm but featherlight touch. I began to wonder: Were Rainey's teachers instilling the right values?

"*Why* do you sit? Do they *make* you sit at school? Do you *have* to sit?" I asked Rainey, my voice increasing in pitch and speed with each query.

But my rapid-fire questions were too much for a three-year-old. My husband, Rob, told me it sounded as if I were saying, "Are your human rights being violated?"

A few weeks later, Rainey suddenly announced at dinner, over stir-fried tofu, "Let's not talk while we're eating."

"Where did you get that idea?" I said, immediately on the case. "Did the teachers say that?"

On this, Rob was also incredulous. "So you can't speak during lunch, Rainey?"

"No. Bei Bei and Mei Mei were talking and the teacher said, 'Be quiet.' Sometimes the teachers are angry to us," Rainey said, as Rob shook his head. Some of the fondest memories from my Texas childhood come from the school lunch table; there, we learned to barter peanut butter for ham sandwiches, negotiate playdates and Friday-night meet-ups, and collect votes for student council elections—and we forged friendships with as much noise as we could muster. Rob also came out of American schools, and I gather he had trouble imagining his own son being subjected to silence over salami (or, in this case, soybeans).

Cultivating a superstar student in China, I'd come to learn, started with impulse control at the earliest age. Jammed into rows alongside his twenty-seven Chinese classmates, planted in a miniature chair, my son had learned to arrange each hand on its corresponding knee, back erect and feet positioned in parallel. He learned never to squirm in his seat or allow a foot out of place, for there was no faster way to draw a teacher's ire. He understood he shouldn't ever touch the student next to him, talk when the teacher is talking, or stand up for water without permission. Above all, he learned the last thing he wanted was to draw attention to himself.

The red star was his reward for sitting mute in a chair, and my son proved a stellar trainee. Though Rainey's toddler talk came out in halting sentences, at home he was clear about communicating one thing: The sticker would not be removed. Rainey brandished the red star at the dinner table, forehead tilted proudly toward the ceiling. He wore it during soccer practice and at a classmate's birthday party. He even refused to peel it off when I tried to wash his face.

"No, Mom—don't touch, don't touch!" he'd yelp, as we readied for bedtime. Off to bed he marched, red sticker intact.

Were we unwittingly engaged in a battle over our son's mind?

"Should we be concerned?" I found myself saying out loud.

"Don't worry about it," my husband would respond, though sometimes I saw his forehead furrow, too.

I couldn't help but worry. Strolling through my Shanghai neighborhood, I observed Chinese children who were proper in public, polite to elders and peers, and orderly on the playground. I sauntered past the neighborhood elementary school at three p.m. on weekdays to see parents and grandparents waiting patiently in a pickup line that snaked around the block: Education here is a family affair. It was no stretch to imagine these children growing up to be disciplined geniuses, respected the world over. But what, if anything, were they giving up?

A journalistic curiosity kicked in as I began to seek answers—by watching closely, asking the right questions, and pursuing experts with more knowledge than I had. As a daily reporter in New York, Minnesota, and California, I'd always defaulted to this approach, and although Chinese society seemed to frown upon independent inquiry, I was driven by a powerful motivating force: parental anxiety.

Ironically, four months after Rob, Rainey, and I had arrived in China in 2010, the country posted some impressive education news: Shanghai teenagers scored tops in the world in math, reading, and science, according to a global student assessment test named PISA (Program for International Student Assessment). In their exam debut, students in my newly adopted city had beat out peers in nearly seventy countries (the United States and the UK finished in the middle of the pack). The result shocked the education policy world. "The Shanghai Secret!" trumpeted the *New York Times*. President Obama declared it a "Sputnik" moment, and the president of Yale marveled in a speech that China was building its own version of the American Ivy League, creating the "largest higher-education sector in the world in merely a decade's time." Meanwhile, media headlines continued to broadcast China's galloping economic juggernaut; not only was China muscling the rest of the world onto the sidelines but was also out-educating the West.

What I was reading in the newspapers didn't exactly sync with my experiences on the ground. Just as I began looking into the school life of these super-achieving Chinese kids, I started to notice troubling signs in our son, such as a habit of obeisance that trickled into other aspects of his life. One day, a classmate's mother asked him if he liked singing. "I don't like singing, but if you want me to do it, I'll do it," he responded. Other days, Rainey would recite Communist Party songs, singing praise unto the "motherland." I would try to convince myself that our home environment was just as important as that of school; meanwhile, I began to watch my son carefully, as if I'd developed a sixth sense especially tuned to subservience and a seventh to brainwashing. I suddenly recalled a conversation I'd had with a European expat friend who had pulled her daughter out of Chinese school. "I did not raise my daughter to be a robot or a people-pleaser," she fumed.

I noticed the Chinese around me had anxieties of their own, but of a different kind. A long-lost but now rediscovered Shanghainese cousin had begun frantic arrangements for his daughter's primary school entrance interviews, enrolling her in a fearsome after-school activity called "Math Olympiad." A high school kid I'd met began marathon preparations for the National College Entrance Exam. A rural Chinese woman who'd babysat Rainey for a year suddenly hightailed it home to Hubei province. The son she'd left behind in the countryside was struggling in the run-up to high school entrance exams, and now he had nowhere to live. "The government is razing my house to clear the way for a building project," the woman said through tears, one of hundreds of millions of Chinese migrants blessed by job opportunity yet simultaneously cursed by the devastation rapid economic change can bring.

Rob and I had headed to China for what seemed like limitless opportunity, but the Chinese themselves seemed uneasy about the change happening around them. I wondered about the contradictions I was sensing: Was the obedience I observed in Rainey part of the Chinese secret to academic success? Was the Chinese education system really pumping out robots or were their students actually getting a superior education? Although the world seemed to hail China's march toward global superstardom, are Chinese methods really what the West should measure itself against, much less emulate?

These questions surfaced again and again, and before long I began taking a pen and paper everywhere I went, jotting notes as I looked for answers. For several years, I trailed young Chinese and talked to teachers, principals, and education experts. I dropped in on schools in the United States and China, and traveled deep into the Chinese countryside to look into reports of devastating poverty and inequality. I pored over research studies and volunteer-taught at a Shanghai kindergarten.

I was certain my reporting would open a peephole onto a massive country that seemed formidable from the outside, yet was quietly struggling to make sense of its newfound standing in the world; and it would also illuminate the best way forward for Rainey in his own schooling.

Early on, I instinctively understood my family would need to bend and be flexible to find our compass (in fact, we'd be required to deliver practice tests before breakfast!). As my journalistic quest began to calm my anxiety as a parent, one important lesson became clear: If we opened our minds, we just might reap the benefits of rearing our child in a second culture, and educating him the Chinese way (while hopefully retaining our Western sense of individuality).

Ours would be a journey that would require grit, more than a few leaps of faith, as well as an outward respect for everything our newfound culture would throw in our direction (including bribery by red-star sticker).

China would have it no other way.

PART I

THE SYSTEM

1

FORCE-FED EGGS

She put it in my mouth, I cried and spit it out,
then she did it again.
—RAINEY

My Chinese uncle told me luck must have been falling from the sky the day I received the news: My son had been accepted into Shanghai's most prestigious kindergarten.

"How did you get Rainey into Soong Qing Ling?" he asked.

In other words, how had I—his American niece—succeeded where he'd come up short? His granddaughter was three, just like Rainey, but she'd been denied a spot. Uncle Kuangguo had been an executive at one of China's first auto supply companies and for much of his life had enjoyed *guanxi,* networks of people he could call on for favors and introductions. He'd had connections of the sort that made the impossible materialize on your doorstep.

"Only luck, I suppose," I told him, as he wrinkled his nose at the floor.

We'd moved to Shanghai from Los Angeles in the summer of 2010, when Rainey was eighteen months old. As we announced plans to move, I was surprised to detect an undercurrent of envy in our American friends' responses, as if we'd hopped a speedboat to China and they'd soon be treading water in our cultural and economic wake. Certainly, America's future seemed increasingly unstable—in the year 2010, the US economy was still spinning inside a recession—while China seemed to be eating the planet's economic lunch. China boasted the world's fastest-growing major economy, the biggest market for autos, the largest number of cellphone users . . . a new superlative seemed to scroll across Western media headlines every other week. Within a decade or so, experts said, China would surpass the United States to become the largest economy in the world.

We'd also been feeling the intense pressure of raising kids in a competitive urban environment. More and more, talk at dinner parties had begun to revolve around how to line up interviews for the few spots available at good American preschools, or the odds of winning that charter school lottery. The decision to relocate seemed easy: Rob had been offered a job as the China correspondent for a US public radio show just as we were closing in on the starting line of an American parenting rat race. "Rainey will be bilingual!" one friend mused with raised brows, as if it had suddenly struck him that a two-language toddler might help with those preschool interviews. "Nanny help is affordable in Asia, I hear," pondered a girlfriend, after she'd battled fiercely with her husband over daycare pickups that week.

Rob and I chose to live in Shanghai's former French Concession, a part of the city center known for its winding streets, from which narrow alleys wind and weave, populated with small shops and cafés and overhung with abundant green foliage. This section of the city was conceded to the French in 1849 but transferred back to China in the 1940s during World War II. In the century in between, the French had made their mark, establishing the concession with now-historic buildings and luxuriant London plane trees. Here, the foreign population enmeshes with its Chinese neighbors, together creating ample demand for such Western delights as French baguettes, gourmet coffee, and fresh Brie, alongside Chinese street delicacies such as scallion pancakes, pork buns, and taro cakes. East meets West, right outside our front door.

We found everything we needed in our new city and made friends quickly. At first, Rainey was more likely to stick chopsticks in his ear than use them to eat, but he adjusted to his new surroundings, too, picking up spoken Mandarin and acquiring a taste for pork dumplings. We hired a Chinese *ayi* named Huangrong to take care of Rainey while Rob and I worked, and they developed a close, playful bond. Rainey spoke Mandarin with Huangrong during the daytime, and English in the evenings with us, and he was soaking up new words in both languages.

"How's Rainey's Chinese coming along?" friends back home would ask with a smidge of envy.

"Great, he's already counting to thirty," I'd boast. "He's perfectly bilingual."

As Rainey's third birthday approached—about a year into our time in Shanghai—he was getting restless at home. A bilingual nursery where he spent time each week changed teachers frequently, and I knew we needed a real preschool with competent administrators. We briefly considered a Western-style school, but tuition for international schools in Shanghai cost as much or more than an Ivy League university, nothing short of insane for a child who couldn't yet wipe his own behind. More than that, we couldn't easily afford it (many American companies covered school fees for their employees, but at the time we were on our own).

Chinese schools are a relative bargain. In China, public schools are managed and funded by an authoritarian government inside a culture that prizes education. It's also common knowledge that Chinese schools are great with discipline, and I reveled in the idea that someone other than I might toil away daily to imbue our son with self-control and a respect for learning, delivering him home at four p.m. every weekday, a model citizen with manners. Rob wanted to raise our son bilingually; he had studied, lived, and worked on four continents, and learning a second and a third language had served him well in his career.

We'd heard stories of Chinese teachers who went overboard with authority and Communist Party propaganda seeping into the later school years, but we couldn't imagine this would be an issue in kindergarten. While we didn't know much about what Rainey's day-to-day experience might be like inside a Chinese school, we liked the big picture.

We weren't alone in that thought. In major Western cities, Chinese nannies had become highly coveted and Mandarin-immersion schools were springing up throughout the suburbs. Friends back home wangled their way onto the board of one such school in hopes of boosting their chances of securing entry for their daughter. China was among the world's largest economies and its mother tongue the most spoken language in the world, and to me it was clear: Being conversant in this culture and country would only become more important.

In Shanghai, we would have a chance to ensconce our child in a real Chinese school—"in China no less," we would harrumph to friends back home.

Our first choice was two blocks from home.

We'd often strolled past the gates of Soong Qing Ling Kindergarten during neighborhood walks. Few Chinese schools looked as stately as this one, with its black-and-gold wrought-iron gates enclosing a sparkling green lawn. A bold government plaque declared the school a "model" kindergarten of the type that all others should aspire to—and in this country, government plaques were a serious business. The school's namesake was the wife of Sun Yat-sen, founder of the Kuomintang, which ruled China until the Communist takeover in 1949. Soong Qing Ling the woman went on to devote much of her life to issues of education and children's welfare, and in today's China she is considered a saint.

The school drew the children of ranking Communist Party officials and families with money and influence. A current Shanghai vice mayor sent his grandchild to Soong Qing Ling, as did a number of other top-ranking city officials. This was significant because the offices of the Shanghai mayor and Party secretary were famous for being rest stops on the highway to national office, producing current president Xi Jinping, former president Jiang Zemin, and former premier Zhu Rongji. Also at the school were the children of tech and security entrepreneurs, IT gurus, and investment magnates. These were families with power, money, and the luxury of choice in a country where, I implicitly understood, options are fewer for those without wealth and connections. Chinese parents who couldn't secure a spot vented their frustrations online: "Many people knock their heads broken against the sky trying to get in," wrote one, using a common Chinese saying. "My child didn't get a seat . . . the world is full of unfair advantages . . . how will my son have a chance?" wrote another. (Clearly, we'd sought a respite from the American parenting rat race, only to land squarely in the Chinese equivalent.)

When I peeked through the school's gates, I imagined the future leaders of China sitting inside its classrooms, being drilled in Mandarin characters, eating noodles and rice, and napping on canvas cots. If China's leadership trust their offspring to this school, I thought, it must certainly be the best early education experience the country has to offer.

Rainey's feelings about Soong Qing Ling were simpler: "I like the playground," he said, tiny hands gripping the black bars of the school gates, peering at the two climbing structures that stretched twenty feet skyward with tiny notches in each post for little feet. The school's classrooms encircled the play structures, and around four p.m. each day young Chinese children spilled out into the green courtyard, their chatter filling the air with merriment.

"*Fang xue le!*" School's out!

"*Mama dao le!*" I see Mom!

"*Hui jia le!*" Time to go home!

There was an element of order as the children covered every part of the lawn, eventually filing through the school's gates with a nod to the security guards.

"*Shu Shu zaijian!*" Goodbye, Uncle, the kids would chirp, using a term reserved for addressing Chinese elders.

Rainey glanced up at me, eyes wide. "Is this my school, Mommy?"

"Not sure yet, sweetheart. We'll see." We'd been preparing him for the transition from days at home: Little boys go to school every day, just like mommies or daddies go to work, we told him.

Rob and I had strategized for months about how to get a spot for Rainey. Most public schools were required to admit students based on where they live, but the government allowed certain schools—particularly model and high-performing schools—to operate by special rules. The Chinese often gained admission to such schools through a complicated web of *guanxi*, but as foreigners we'd need to find an alternative way in. During the year before Rainey would enter *xiaoban*—the Small Class grade level that is officially the first year of kindergarten—we called the main office again and again. Rob and I took turns, carefully spacing out our calls by weeks, hoping they wouldn't remember our voices from previous failed inquiries.

Each call was met with the same detached voice, giving us some variation of the same answer: "Our classes are full. There's no reason to call again. There's no hope for a spot." We asked to speak with the principal and were always told, "She's away at a meeting." One time, I managed to get her name: Zhang Yuanzhang, or Principal Zhang.

We always had a backup in Shanghai's international schools, but as we neared summertime, I figured I had nothing to lose. One more time I walked over to the school entrance. I peered into the guardhouse, enclosed in glass and set just behind the black iron entrance gates, in perfect position to detect intruders.

In moments such as these, I'm always aware that I'm a special kind of foreigner. Mandarin was the language of my childhood home, as my parents had emigrated with their families from China to the United States via Taiwan decades before, but I was born and raised in America. Thus, I speak Mandarin with a hint of a Texas drawl and a vocabulary of California-speak I cultivated as a young adult. To most mainland

Chinese, I appeared to be one of them until I opened my mouth, and they'd quickly identify me as someone deserving of unique disdain: an alien cousin who spoke oddly and had foreign mannerisms.

I waved at the guardhouse glass. One of two men sitting inside ventured out.

"What do you want?" he barked through the gates. He was a short man, coming up only to my shoulder, and he wore thick glasses.

"Is Zhang Yuanzhang in?" I said.

"What are you inquiring about?"

"Just wondering about a spot for my son." He looked me over and then shook his head.

"It's full, it's always full," he said. "There's no use in asking."

"May I speak with her? I'd just like to introduce myself."

"She's in a meeting," the guard responded. He turned and stepped back into the glass-enclosed house.

"When is a better time to come back?" I called after him, but he ignored me. Later, at home, Rob and I regrouped.

"Should I go?" Rob asked.

"I think you might have to," I replied.

"Yup," Rob said, looking at me meaningfully. That was the reality in China: Sometimes you needed to send over the "foreigner." I spent my first year in China bemoaning this reality, as a strong-headed woman who managed the finances in her marriage, but here was a culture with an ingrained hierarchy about who matters and who doesn't. Caucasian Americans trump Chinese Americans, and on top of that, Rob was a fluent Mandarin speaker who'd spent time in rural China. He was a source of unusual fascination for the Chinese, and he typically managed to sneak in a request while they wallowed in wonderment at this blond, blue-eyed foreigner who spoke their language.

Rob walked over to the school gates before work one day and triumphantly reported that evening that he'd managed to get the guard to call a teacher out to meet him.

"She was friendly," Rob reported. "She said, 'You speak Chinese so well!' Then she wrote Rainey's name and passport number down on a piece of paper."

"Did she write it down in some kind of notebook?" I prodded. "Did it look like a waiting list?"

"No . . . it was a scrap of paper. The size of a Post-it," Rob said, with some hesitation.

"She didn't invite you inside?" I pressed, my pride assuaged by the fact that he, too, had failed to penetrate the gates.

"Nope." Rob raised his eyebrows—he knew exactly what was going on inside my head. Even so, we rejoiced. Somewhere inside that formidable institution, our son's name was sitting on a slip of paper that might eventually find its way to Principal Zhang.

Now we could only wait.

A month later a call came to my mobile phone. The vice principal introduced herself as Xi. "You're lucky," she said brusquely. "We are allowed to add a few spots to the *xiaoban* grade level. Has Rainey found a school yet?"

"No, we'd love for him to go to Soong Qing Ling," I said quickly.

"Bring him over next week."

On the day of our "interview," I dressed Rainey in a plaid shirt and corduroy pants and served his favorite breakfast of oatmeal and apples. Rob was traveling for work, and my father happened to be in town for a visit, so I decided to bring him along as backup.

I planned to present to Principal Zhang the portrait of an ideal Soong Qing Ling family: captivatingly cute toddler, engaging Mandarin-speaking foreigner parent, and one talkative grandfather with deep roots in the country. Rainey said "*Nihao*" when he was to supposed to say hello, I hit my Mandarin tones perfectly, and my father became fast friends with Vice Principal Xi, who marveled that here was a Shanghai-born Chinese man who raised a family in the United States, while decades later his American-born daughter was rearing a family in the motherland!

Whether by chance or circumstance, a family had gone on a long journey and returned to the bosom of the motherland. Now its members were kowtowing to Vice Principal Xi. "And Rainey has the opportunity to attend Soong Qing Ling," Xi said, perhaps pondering the smallness of the globe or the circular nature of life.

"Yes, we hope so!" my father smiled.

"*Qian ren zhong shu, hou ren cheng liang*," Xi remarked, shaking her head. "One generation plants the trees, another gets the shade." That was it: the moment I knew we were in. A Chinese proverb uttered in conversation indicated wonderment, camaraderie, and acceptance all rolled into one; it was always a splendid surprise, like a tissue-wrapped moon cake dropping suddenly out of the sky. Principal Zhang, who had hovered nearby during most of our meeting, delivered a short nod at Xi before stepping out of the room.

"So Rainey's in?" my father asked Xi.

"Yes." Xi gave a firm nod.

At home, Rob and I marveled at our good fortune. Rainey had a spot in school and we couldn't be more thrilled.

* * *

IT WAS RAINEY'S first day of school. Rob and I made our way through the crowds that thronged the streets of the French Concession on this lovely fall morning. Between us, like a tiny bud on a daisy chain, was Rainey, each hand clasping one of ours. As Rob and I took a step forward, Rainey pulled us backward, tugging with all his might.

Suddenly, Rainey stopped. "I'm going to cry," he announced.

"It's okay to cry," I assured him, dragging him forward. Slowed by our hand-holding, our little ensemble weaved its way north up the crowded sidewalks of the street in front of our complex and waited out a red light as taxis, sedans, and scooters choked and gasped on the morning commute.

"Let's not cut through the hospital," Rob called out. We passed a bustling hospital complex, whose outpatient department alone serves more than four million patients a year, and veered left onto the main road. As we approached the school, we stepped in front of a Ferrari and around the tail of a BMW with a driver inside, assisting with morning drop-offs. In the Soong Qing Ling crowd, Rob and I were distinctly middle class, but we were foreign and that was our mark. Indeed, Rainey was the perfect manifestation of West and East. He had my husband's lanky figure and high-bridged Caucasian nose, and the dark hair and black-brown eyes of my Chinese heritage. His round eyes were so large, they overtook most of the real estate on his face; the effect was that he looked continually awestruck. Old Chinese men and young professional women alike would stare down at him, exclaiming, "Look at the little foreigner! He's a handsome little guy!"

We streamed through the wrought-iron gates along with other sets of parents and grandparents, as the elders issued missives in Chinese: "Behave. Listen to the teacher. Eat your vegetables."

Soong Qing Ling divided its classes into four grade levels—"Caring," "Small," "Middle," and "Big"—and further divided each level into classroom numbers. That made twenty classes total, spanning the ages of two to six years old. Rainey's home for the next year would be Small Class No. 4.

Parents and grandparents crowded the classroom doorway, packed four or five deep. Over their heads I saw the face of Teacher Chen, master instructor of Small Class No. 4. She had yellowed teeth with black marks where they touched, and, despite her smile, my gaze always seemed to home in on those dark spots. I instinctively gathered there was nothing small about her temper.

Today, Chen was friendly, patting children's heads and fluidly greeting parents in either Mandarin or Shanghainese: "Welcome to school! Did you have a nice summer?" She intuitively knew which language to use. The children seemed calm and compliant, issuing a nod or chirping a hello.

In China, you don't so much push your way forward as let the crowd carry you. As we were slowly transported to the door frame, Rainey gripped my hand tighter and tighter. Parents peeled away and we finally found ourselves at the door, facing Teacher Chen. I nodded at her, adjusted Rainey's shirt, and gave him a reassuring pat on his head. "We love you, Rainey. We'll see you after school."

"No! Don't leave me," Rainey said in English, glancing up at me frantically as Rob and I pushed back toward the stairwell. "Don't leave me *here*." Tears began flowing. I sidestepped over to the exit, and in a flash Rainey went horizontal, throwing himself at my ankles.

"We'll see you later today," I said, looking down at the back of Rainey's head, trying to stay calm. "You'll have fun at school—see all the toys!" I looked around. Pale green streamers hung from the ceiling at evenly spaced intervals, and a row of paper plates with flower petal cutouts greeted us along the back wall. Color drawings from the previous year's class were still up; I noticed that all the apple trees were colored in the same way, with scribbles in shades of green, brown, and red.

"Go have a snack. Be a good boy. *Tinghua*," said Teacher Chen. "Listen to what I say." It was the listen-and-obey command that would become a theme of Rainey's time here.

"Noooo!!!" Rainey sobbed, speaking to my ankles as Chinese parents looked on with great interest. It seemed their children were silent, standing tall and filing into the classroom, spines erect.

"Don't go. Don't leave me!" Rainey screamed. Rob leaned over, detached Rainey from my ankles, and carried him into the classroom. He deposited our thirty-pound boy near the play kitchen, turned, and sprinted toward the doorway, next to which several children sat at tables, quietly drinking milk and chomping sugar cookies—cookies at eight a.m.?—and glancing our way with interest.

Rob and I pushed through the crowd of parents and children and ran down the stairs; we stopped once we were safely out of sight. I turned my ears in the direction of the classroom door, hoping to hear silence: the stealth listening familiar to any parent doing a daycare or school drop-off.

Rainey was still wailing. We listened for about thirty seconds, and I suddenly noticed I was grinding my teeth. Rob turned to me.

"Welcome to Soong Qing Ling," he said, with an uncertain smile. He grabbed my hand and we made our way toward the school's gates.

The second day of school, Rainey came home with a story. Four times, he found egg in his mouth. He hadn't placed it there himself; instead, his most hated food made its way past his teeth by the hand of the fearsome Teacher Chen.

"She put it *there*," Rainey told me, mouth wide, finger pointing inside. Then what happened? I asked.

"I cried and spit it out."

Then what?

"She did it again," Rainey said. We were making our way home along a crowded sidewalk jammed with people and lined with fruit and vegetable vendors. Rainey jabbed an elbow into the bottom of an elderly Chinese woman walking before us, pushed firmly, and stepped into the void he'd created. Our little boy could throw elbows like the natives.

And? I prodded.

"I cried and spit it out again," he said. I coaxed more of the story out of him; Teacher Chen had put egg in his mouth four times, and the last time, he swallowed. I absorbed the news of my child being force-fed in school, the clamor of China surrounding us: the bleats of a vendor hawking tomatoes on a three-wheeled cart, the creak of a turnstile halfway down a neighborhood lane, the sounds of taxi horns. At home, battles over food ended in screaming and flailing. Rainey would rather go hungry than try something new. The last time I'd presented him with steamed kale (which I arranged in dinosaur shape on his plate), he thrashed so violently that he chipped a tooth against the floor.

"Why did you eat the egg?" I asked.

"I don't want to talk anymore."

Later, I texted an American girlfriend, thumbs flying over my phone. "I think Rainey was force-fed eggs in school."

My friend responded without missing a beat. "I think the severe discipline of the Chinese teaching style is oppressive. . . . You are not going to find Mary Poppins here. But forcing a smart, freethinking child to eat an egg is disturbing." Her children attended an international school in Shanghai.

I resolved to ask Teacher Chen about it at pickup time. I didn't entirely trust my three-year-old's account of what happened, and I thought it would be a great time to introduce myself as an engaged and involved parent. The next day, I lined up with the crowd—mostly mothers and grandparents—outside the classroom door and strained to lift my head above the throng. I got a glimpse of thirty children sitting in pint-size chairs in a perfect U-shaped formation around the teacher. Rainey was sitting—sitting!—in the third chair from one end. He rarely sat at home, and I saw here that it wasn't easy for him. Every appendage was in motion: elbows, hands, legs, feet. He looked like a wiggly caterpillar with its middle nailed down and the rest flailing wildly, but still he kept his seat. Craning his neck toward the door, along with every other child awaiting pickup, Rainey finally spotted me. The teacher saw me also and called his name. Rainey sprang from his seat and ran over to me. One by one the other children's names were called, parents peeled away, and finally I could step over to Teacher Chen. Rainey clutched my hand.

"Good afternoon, Teacher Chen," I said. "We're excited to start the new year."

"Good, good," Chen responded.

"How's Rainey doing?"

"Good," Chen said, nodding.

"I wanted to ask something. Rainey doesn't eat eggs at home, but he told me he ate them at school. Did someone put eggs in Rainey's mouth?"

"Yes," she answered, without a smidgen of defensiveness. My heart jumped, and any plan to ease into this conversation evaporated. I forged ahead.

"Really? Who?"

"I did," Chen said.

"Oh! Ah! We prefer you don't make Rainey eat foods he doesn't like. We foreigners don't . . . use these methods," I said, as Teacher Chen tapped her foot and glanced past me. "We don't use such methods of force in America," I repeated.

"Oh? How do you do it?" she clucked, looking down at Rainey.

"We explain that eating egg is good for them, that the nutrients help build strong bones and teeth and helps with eyesight," I said, words speeding up as I tried to sound authoritative. "We motivate them to *choose* to eat eggs. We trust them with the decision."

"Does it work?" Chen said.

An image of Rainey's chipped tooth flashed before my eyes. "Well . . . not always," I admitted.

Chen nodded. "Rainey needs to eat eggs. We think eggs are good nutrition, and all young children must eat them." She took one last look at Rainey and then clicked away.

* * *

THERE'S A LITTLE MAN who resides inside the head of every Chinese man and woman, whether they know it or not. He governs how they find their mates, what they look for in jobs, how they treat their parents, and how they educate their children.

His name is Confucius and he lived 2,500 years ago. He was both a teacher and a philosopher. You mention his name—*Kong zi*—to most ordinary Chinese, and they nod in respect and then change the subject, as if he's some great-great-granduncle they know they're supposed to revere but don't have much to say about.

Confucius believed that the purpose of education was to shape every person into a "harmonious" member of society, and harmony was more easily maintained if everyone knew his proper place. So Confucius spent a lot of time and attention delineating relationships. In Confucius's world, a wife always obeys her husband, a subject never challenges his emperor, a young brother heeds the older brother, and a son does what his father requires on a daily basis. "Let the ruler be a ruler, the subject a subject, the father a father, the son a son," Confucius said, according to the Analects, a classic collection of ideas attributed to him and his followers. Confucius staked his entire philosophical pantheon on the concept of top-down authority and bottom-up obedience.

I remember these lessons from my own upbringing. My father grew up in Asia and brought his Chinese ways to America, where he met and married my mother—like him a fellow Chinese immigrant—and raised my sister and me. My father was nothing if not authoritarian, and I had a rebellious streak, setting the stage for many clashes during my childhood. In our suburban Houston mid-century house sat a preserved portion of a tree that had grown in the yard of my father's childhood home in Taiwan. Knotted where branches used to grow and slicked over with polyurethane, the section of trunk served as a gnarled, daily reminder of ancient traditions.

I was constantly overstepping boundaries. "I don't want to do homework tonight, I'm tired of chemistry. You make me study too hard," I'd say to my father, whom I called *Ba*, the Chinese word for father.

"A daughter is not supposed to talk to her father like that," he'd retort, lips quivering with rage.

As a teenager, I never knew who had told him daughters don't talk like that, but now I know it was Confucius whispering his wisdom across the generations and reaching down into our lives. Another time, I coolly informed my father that I was plotting to audition for the drill team, a venerable institution at a top high school—Texas state champs in '78—inside a state that worships football. Unfortunately, football and dance don't register as a worthy pastime on the Chinese scale.

"Dance is not an honors elective," Ba said. "It will bring down your grade point average."

"I don't care. I'm going for it anyway," I told him.

"Oh?" Ba said. "You can just decide like that?"

"Yeah. I'm going to do it, and there's nothing you can do about it."

I remember there was nothing more satisfying—and terrifying—than challenging my father's authority. That was also me butting up against Confucius.

Every year, my mother and aunt—my father's sister—would spend days preparing dishes for our annual Chinese New Year feast. They'd shell shrimp, chop pork, braise mushrooms, boil winter melons for soup, roast ducks, wrap dumplings, and on and on. At the end of the long labor, we would place the dishes—sometimes two dozen in all—on a table alongside burning incense as a way to honor our ancestors. After an hour or two, the smoky, sweet smell of incense would pervade the house, while the food turned limp, with stir-fry oil forming shiny pools on the plates. We'd have to reheat each dish one by one in the microwave before we could finally sit down to enjoy our feast. The year I was ten, I announced that this ritual not only seemed inefficient but was a grand waste of feminine labor.

"We must honor our ancestors," my father countered, launching into a speech about tradition. The lecture, a familiar one by this time, would take fifteen minutes but boiled down to the promise that misfortune would befall those who didn't honor their ancestors properly. According to the Confucian canon, an event that seemed like bad luck—an accident, a lost job—was really an ancestor's revenge. And so the women of the family continued the annual shell-chop-braise-boil-roast-wrap marathon, and we continued to let the food sit on a table for hours afterward.

I was a terrible practitioner of filial piety. This is one of the concepts Confucius espoused most heavily, and it called for children to respect their elders as well as folks who'd given rise to their bloodline but were long gone from this earth. While in the West, elders often retire to nursing homes when health turns frail, the Chinese buy homes with a spare bedroom for Grandma and spend days preparing feasts to honor her after she's gone.

Ancient paintings and verse illustrate how the virtue of filial piety knows no bounds: A young boy encourages mosquitoes to suck his blood so that his parents will remain undisturbed while they slept; a man buries his infant son alive to free up scarce foodstuffs for his ailing mother; a man tastes his father's stool to diagnose a terminal illness and then offers up his life in exchange. My favorite is about a woman and her mother-in-law, a figure much maligned in Western culture. Not so in China: In one tale, a woman cut out a piece of her own liver to boil as a base for broth to cure her severely ill mother-in-law.

In today's China, filial piety often manifests itself in the form of educational achievement. Performing well in school is the ultimate way to respect your parents, since good grades and test scores are the path to financial stability, and the ability to provide for your parents and buy a bigger house—with a room for Grandma. The principle of obedience and respect extends naturally to teachers, of course, as children are "required to respect the teacher's authority without preconditions," wrote two Western academics in a 1996 study of Chinese kindergarten curricula.

Parents are also required to fall into step. And so, I had crossed a line with Teacher Chen. I had failed yet again to be a proper practitioner of Confucianism.

I sensed it the moment she chose to end our conversation about eggs with a simple maneuver: She walked away from me.

Over the following weeks she no longer engaged me, although I could see her glancing my way amid the scrum of parents at pickup. On the day she asked me to step into the coatroom, I realized she'd been waiting for an opportunity to talk privately.

"Let me bother you for a second," Chen said, using a common nicety.

"Sure, is there a problem?"

"No problem, but . . . I want you to refrain from questioning my methods in front of Rainey," she said.

"Oh," I said.

"Yes, it is better that the children think we are in agreement about everything," she said.

"Oh!" I said.

"If you have a different opinion from me, then you can talk to me privately," she said. "But in front of the children you should say 'Teacher is right, and Mom will do things the same way,' okay?" she said.

"Oh . . . okay," I stammered. "I didn't mean . . . I just want him to be happy in school."

"Happy? He is happy," she said, with a firm nod. "And at home it's best also if you don't talk about the teachers when Rainey's in the room." Teacher Chen's mouth was upturned, in a taut smile of forced courtesy.

Before I could reply, she admonished me once more. "Don't make the children feel Mom's opinion conflicts with the teacher's."

I nodded and quickly stepped out of the coatroom. I'd been scolded by Teacher Chen!

On my way home with Rainey, my mind raced to process the exchange. What I'd viewed as a simple request of Teacher Chen had been taken as an affront to her authority. The glimpse I'd had of Rainey sitting in a chair at pickup time, waiting for a teacher's permission to rise, suddenly seemed foreboding.

Snippets of conversations I'd had with other Chinese parents filtered into my head—they suddenly had context. A Chinese mother had told me once she was afraid to ask her son's teacher about anything, for fear her child would suffer consequences in the classroom. Poor woman, I'd thought back then. Imagine that! Another parent had told me she spent weeks planning the gifts she'd bestow on her daughter's master teacher. Western skin-care or luxury items were great, she said, enthusiastically ticking off brand names with a Chinese accent: Louis Vuitton, Prada, L'Occitane, Clinique, Godiva. She'd been tipping me off, I now realized. I'd failed these tests, and to top that off I'd challenged a teacher in front of a child. A Chinese early childhood education professor would later tell me that the worst thing for a teacher is to lose face in front of children. "Westerners may feel happy when children challenge them, but in our traditional culture we don't think life is supposed to be like this," he said.

The following weeks Chen avoided me as much as possible. When I approached she was always engaged with another parent or leaning over to talk to a child. Shanghai built its subways in just a few years, and a new building seemed to pop up in our neighborhood every month. Shanghai Tower was rising on the horizon, steel piece by steel piece, and when completed it would be the world's tallest skyscraper, boasting the fastest elevator in existence. The country's economy was growing at unparalleled rates. But amid all that change, some traditions were too ingrained to shift even an inch: You must be devoted to your elders, and you never questioned your teachers to their face. Today's rules of societal conduct had been cemented two and a half millennia ago, and I had done the unthinkable.

Rainey and I were in trouble.

Over the next week, I began to watch others in the pickup line, keeping my head down as I approached the classroom door. Some yapped with Teacher Chen and Associate Teacher Cai in their native Shanghainese—a language I didn't understand or speak—and others spoke in Mandarin about the children's artwork or after-school activities. The parents never seemed to voice any concerns or brave any inquiries about what the children were doing in school. The only questions I heard were open-ended and nonconfrontational, innocuous: Did Nong Nong fall asleep quickly today at nap time? Did Mei Mei eat her lunch?

All exchanges ended in some sort of nicety by the parents: "*Laoshi, Xinku le*!" Teacher, you work too hard! What a great job you're doing! "*Taibang*!" Too amazing! Excellent!

Okay, I get it, I told myself. Keep it light. Offer compliments.

One day at pickup, Teacher Chen spoke to me: "Rainey eats eggs." Her delivery was flat, as if she were ticking off a grade on a list of subjects. Her tone of voice clearly denoted B-minus, as in, "The process was labored, but the goal was eventually reached."

"Taibang le!" I cried. "Excellent!" Inside, my stomach was inside my throat. Eggs had taken on a new meaning in my life; they were no longer simply a good source of protein. I was anxious about eggs, and fearful of the methodology employed to get my son to swallow them.

Safe inside our apartment walls, Rainey continued to refuse eggs, whether scrambled, fried, poached, or boiled. How did Chen manage it? And did he eat them willingly? By his own hand? These thoughts kept rolling over in my head, until I decided I needed to see Rainey eat an egg for myself. One weekend morning, I cajoled Rainey to have an egg, tossing in a helping of carbs and offering a bribe for good measure.

"Rainey, if you eat this French toast you can watch *Kungfu Panda* after breakfast," I said.

"What's French toast? Does it have egg in it?" Egg clearly had special meaning for Rainey, too.

"Well . . . French toast is bread. With a little egg," I ventured. "But . . . mostly bread."

Rainey glanced at the egg-coated creation I'd placed on a plate, thought about kungfu-kicking pandas, and nodded. He asked for a cup of water, and I installed myself in a chair about ten feet away, pretending to read a book.

Rainey placed the plastic water cup at the two o'clock position, next to his plate. He looked at the arrangement for about half a minute. Then, deliberately, he started. His little fingers tore the French toast into small pieces, which he scattered about the plate. He eyed the pieces. He took a deep breath. He chose one and placed it in his mouth.

Morsel in, he moved quickly, scooping up the cup of water and tipping it into his mouth, flushing the bit of French toast down his throat like a dinghy pressed through a tunnel by a tidal wave. He paused, took another breath, and repeated the process. I refilled his cup when empty. He was determined, and he didn't pause to speak or smile. Fifteen minutes and three cups of water later, the French toast had disappeared.

I heard the voice of every American pediatric and nutritional expert from Mayo Clinic to Dr. Sears: Don't ignite food battles. Don't force food on children. Don't let mealtime become a source of anxiety. Food should be enjoyed, lest disordered eating develop in later life.

It was safe to say I had accomplished all those no-no's in magnificent form. My promise of *Kungfu Panda* was also a Western no-no—never reward food consumption. Yet I was amazed. I'd never suspected my three-year-old was capable of such a display of resolve. I barely recognized my tantrum-throwing, food-tossing, chipped-tooth little boy in this creature who'd willed himself to complete a task he did not enjoy. The culture reinforced the notion that teacher knows best, but did the end justify the means?

After the plate was empty Rainey settled in front of the television, *Kungfu Panda* cartwheeling across the screen.

"Do the teachers watch you eat?" I asked. Rainey paused, considering my query.

"I'll tell you if you let me watch *Thomas the Train*, too."

2

A FAMILY AFFAIR

We don't pick children—we pick parents. Yesterday, I met a parent who purchased two flutes. One flute for herself, so she could practice alongside her child. I like that kind of parent.

—PRINCIPAL ZHANG

A few months into Rainey's first year of kindergarten, my Chinese acquaintance Wu Ming Wei brought her son Little Hao over for a playdate.

"You are very free in allowing your son to play," Ming told me. It was an insult, delivered the euphemistic Chinese way. She watched Rainey chase after a ball that had rolled under a table. On the way, my son hopped on a chair and leaped off the armrest with arms flapping, as if celebrating that something had gone awry in his little world.

Ming had all the accouterments of a middle-class city parent: a stable job as a doctor; two sons birthed legally through an exception to China's one-child policy; an apartment with a room for Grandma, who did indeed live in the room; an address a block away from a well-regarded public kindergarten; and plans for her children.

I considered Ming something of an informal education expert. Clearly, she didn't think the same of me.

I glanced over at her boy, sitting cross-legged in front of a pile of Legos. Ming had arranged everything for him, pulling out the pieces, organizing them on the floor, and even handing him the Lego he should start building with.

"We do have rules," I told Ming, suddenly defensive. "But jumping off a chair and crawling under the table isn't dangerous. I don't see anything wrong with it."

Ming pondered this thought. "You're letting him explore—that is a luxury," she said, with a tinge of envy, as she watched Rainey scamper out from under a table, ball in hand. "Foreigners are more free."

Meanwhile, Little Hao was building a structure with intense concentration.

Rainey had spurts of focus, but he rarely sat still and never for fifteen minutes at a time. A child in motion is a whirlwind, moving forward and backward, jerking to satisfy any kind of impulse he might have. It was quite a feat to still all of this activity. How does one cultivate focus in a child so young? I wondered. Is that even a good thing?

"I *am* a foreigner but our children are in the same education system," I told Ming. "Perhaps because we are more free with him, Rainey is having trouble adjusting to school."

Ming nodded, as if I'd confirmed some kind of secret suspicion she'd held about me. "What *wai ke* is he taking?" she asked.

"*Wai ke?*" I echoed, almost to myself. Outside classes? Already? I looked at our sons—hers sitting and stacking, and mine sprinting circles around the dining-room table. Surely she must know what my answer would be.

"I haven't thought about it," I said. "What is Xiao Hao taking?"

"English, math, and *pinyin*," Ming answered. *Pinyin* is the phonetic representation of Chinese characters. Learn the *pinyin* for any character, and you'd immediately know how to pronounce it. For example, *haizi* is the pinyin representation for 孩子, which means "child." *Xuexiao* is pinyin for 学校, which means "school."

I recalled the time another parent lectured me about extracurriculars.

Gregory Yao's daughter was only five years old, but she was already taking eight classes a week, including math cram classes and the "early MBA," which trained babies as young as four months in "six core areas," including leadership and global vision, according to one provider.

"Why? Why start so early?" I'd asked Yao, who exuded the pressure—as well as the internal conflict—of the modern Chinese parent, in his zip-up nylon jacket and bowed shoulders.

"The four best elementary schools in Shanghai take one child out of every three or four hundred who apply," Yao told me. "It's like a chain effect. You can only get into a good college if you're a graduate of a good high school, a good middle school, a good primary school, and a good kindergarten," Yao said, squinting through rimless glasses. "The competition starts early."

To me, the parenting conundrum was obvious. College competitiveness is one thing, as a student has had seventeen years to distinguish herself through leadership, activities, and grades. Stacking a very young child up against her peers after only five or six years of existence on earth is quite another.

How could a child possibly hope to distinguish herself?

In Yao's mind, the answer was simple: standing a head above the rest in something, anything. Yao could increase his daughter's chances of success by shaping her into the best multiplication-table-reciting, calligraphy-drawing, piano-playing student money could possibly buy, all before she could cut up an apple by herself. All told, Yao spent nearly $1,000 a month on eight classes a week—nearly all of his disposable income went toward educating his child. Chinese parents actually spend more on their kindergartners than on their high schoolers—about a third more—to launch their children with the right foot forward.

I glanced at Yao, whom I'd met while interviewing parents for a magazine story about Shanghai education. He had a habit of pressing together his thumb and forefinger, as if he were trying to gauge how many bills might be in a stack of cash, and just standing near him made me nervous. If my little boy was to stay in the Shanghai system, we would surely find ourselves climbing this ladder system of study-pass-advance, and Rainey's competitors on the scramble would be children like Yao's daughter. And these kids were learning math and English outside of school starting at the age of three and younger. In conversation, a parent once mentioned to me that eighteen million babies were born in China every year; her anxiety suggested she imagined that a mound of infants equal to the combined populations of New York City and London would soon rise from their cribs, prepared to compete with her son for school spots and jobs.

My own father had clear expectations for my time: During high school, I was programmed as tightly as a cable box with a thousand channels: Advanced Placement classes, Academic Decathlon, SAT summer prep, Sunday Chinese school, and a few other parent-prescribed activities I've blocked from memory (I'm pretty sure they involved a No. 2 pencil). There was more: Grades should be perfect, dating ignored until college, and dance class and sports strictly elective. In my last few years at home, my father and I fought viciously over the right to direct my future. Was my life his to dictate or mine to own? Ours was a classic story of Chinese expectations meets American culture and a strong-willed personality. Yet, my father chalked up a win for his side the day I got into Stanford University.

As a parent, I like to think I harbor my father's high expectations, only with a silky-soft grip and a dose of empathy. I wanted Rainey to express himself, explore hobbies, and forge his own path in a way I never did as a child. In other words, I have plans for my little boy, but I wasn't yet ready to tiger-parent a toddler. Rainey most certainly wasn't studying *pinyin* like Ming's son or taking Genius classes like Yao's daughter. Rob and I had registered Rainey in a weekly soccer league, but other than that, he loafed around on weekends.

Ming clearly thought this was dangerous, and she felt compelled to explain the risk of inaction.

"It is good for kids to be free now, but in China, eventually all students must walk a very narrow road."

★ ★ ★

THIS KIND OF anxiety spared few parents in China, regardless of geography or class, and expectations for a child's behavior were correspondingly strict. I remember this lesson from my early interviews for an ayi—a nanny to watch Rainey while I worked. Pronounced "ah-yee," the word literally translates as auntie. In our home ayi would come to mean housekeeper, cook, babysitter, tutor, and friend.

During my first month in Shanghai, I'd called on an agent who had given herself the English name Carol. In a city of twenty-six million people, I'd need a middleman. Carol told me she had a database of men and women, categorized by height, weight, hometown, skills, experience, and salary requirements.

"My ayis can shop for groceries, make dinner, clean your house, and watch your child," she said. For the equivalent of four US dollars an hour and a few meals?

"I think that will be fine," I told Carol.

Most ayis are one of hundreds of millions of Chinese men and women who filter into big cities from the countryside, drawn by salaries double and triple what they could make back home. Nearly always the decision to *chu qu*—go out for work—is fueled by the need to support a child's living expenses and schooling back home. (Countless individual decisions, big and small, are made in the name of education.) China's rural-to-urban migration is the largest mass movement of humans the world has ever seen—about 350 million over the past few decades—and I'd be helping to power the pilgrimage by creating a job opening in my Shanghai home.

"My ayis are *shou jiao hen gan jing*," Carol continued, which literally translates as "hands, feet, very clean." To have *un*clean hands and feet is to be a petty thief.

"Give my ayis five hundred yuan to buy groceries for your house, and she'll buy five hundred yuan's worth," Carol promised me. "She won't spend four hundred and fifty yuan and put the change in her pocket. Where are you from?"

"America—I'm American."

"American!" she exclaimed. "American families are good to work for! Ayis like to work for American families. And Canadian. And British."

"Why do ayis like to work for Americans?" I asked.

Carol ignored me. "Ayis don't like Germans. Germans refuse to negotiate on salary. Ayis don't like to work for Spanish, either. Spanish are messy. Unrestrained. Always late. Singaporeans have one washing machine per family member—too many machines for ayis to manage." I chuckled. I often saw Chinese generalizing about a people or culture they'd experienced only in passing.

"Hong Kong families don't like ayis to sit on their couch," Carol continued. "French are arrogant, plus they like their ayis young and pretty—most ayis don't qualify. Indians are vegetarian, so ayis who work for Indian families are always hungry."

"What do you mean?" I said.

"Starving. Ayis will call me and say there's nothing to eat but one tomato, a carrot, one potato. And some curry paste," Carol said. I suppose in a country so carnivorous that fifty-four million *tons* of pork are consumed every year, a meat-free workplace is a downer.

I repeated my question. "Why do ayis like to work for Americans?"

"Because an American will give Ayi medicine if she's sick. She won't send Ayi shopping in the rain. She'll invite Ayi to eat at the dinner table with the family, instead of in the kitchen. She follows Ayi around to ask, 'Are you happy, are you happy?'" Clearly, in Carol's eyes, the American mother was a neurotic boss concerned about human rights and labor conditions.

"If I have an American family who needs an ayi," Carol said, "I'll walk into my dormitory and every ayi will raise her hand. 'I'll go, I'll go!'"

Carol said she would head over in a few hours, and soon enough she appeared on my doorstep with another agent and three smiling ayi candidates. Carol wore a black patent-leather trench coat and shiny black pumps, attire that seemed to advertise: Hire one of my ayis and you'll have the time to primp like me.

The Chinese women filed into my dining room, and each settled into a seat around our table. I took a chair. Five pairs of eyes the color of black tea focused on mine.

"You can start the interviews now," Carol said. I surveyed the army of strangers seated where I eat breakfast.

"Now? Here?" I asked.

"Go ahead. Ask their names and experience," Carol prodded. I started with the woman on my left.

Tang Ayi took pride in the fact that she'd worked for only three families in the eight years she'd been in Shanghai—she was loyal—and she was pretty, with a gentle manner about her. She looked over at Rainey, who was busy maneuvering trains about the living room floor, and smiled.

Wu Ayi was from Anhui province, which bordered Shanghai. She made sure to announce that tap water was not for drinking and that she'd been a chef back in her hometown. I imagined the meals she could make.

Hu Ayi was forty-four years old from Fujian province, near the ocean. It was said that Fujian had the cleanest air in all of China, but Hu Ayi looked as if she lived in a chimney. She'd likely just stepped off a dusty, multiday journey from the countryside.

All the women were eager to work. Each explained her experience, but I was confused.

"Which do you like?" Carol asked, as if we were discussing the purchase of a gerbil at a pet store. The ayis sat waiting, both winner and losers sitting together in my home. It occurred to me that the situation perfectly mirrored China's labor force: Population is large. People are dispensable. One doesn't suit, a replacement is always at hand. There was never any need to gloss over this fact of life in China, and I felt anxious knowing my choice would ultimately help support a family's child back home.

"How do I choose?" I finally asked.

"I'd invite each ayi to your home for a day," Carol said. "Watch how they wash the walls. Do they put chemicals in the water, or do they just pass over surfaces with a wet cloth? How do they iron? Do they chop vegetables neatly? And you need to see how they massage the baby."

"Massage the baby?"

"Yes," Carol said. "The French think this is very important."

"Well, if anyone will massage the baby," I said, "it will be me."

Carol nodded. "Why don't you try each one for a day. Which would you like to start with? Pick one." With her chin, she traced a circle around the table.

This was too much. "Could they wait outside?" I asked. Carol shooed the women out the front door and came back to the table, ready to talk business.

"The first one you could get for two thousand RMB* a month," she said. That was about $300, which was a nice monthly Chinese salary in the year 2010. "The second you'd have to pay twenty-five hundred, and the third one will want twenty-three hundred. My fee is forty percent of a month's salary."

The next day, Tang Ayi showed up at eight a.m., ready to work. She positioned herself next to Rainey, who was sitting at the dining-room table, spooning oatmeal into his mouth.

"Hello, little boy," Tang Ayi said. The little boy did not respond.

"His name's Rainey," I told Tang Ayi.

"Rainey, what are you eating?" Tang said, staring him down intently. Still, Rainey did not respond, and apparently that was simply too much. Tang would make him acknowledge her, and she abruptly removed the spoon from his hand and tried to force it into his mouth. Rainey clambered down from his chair and ran over to me.

* RMB stands for renminbi, the official name of Chinese currency; the colloquial name is yuan.

Tang followed him, spoon outstretched. "Little boy, come eat this porridge now," she said.

"You can call him Rainey," I said as he clung to my arm.

"Rainey, come over to me," Tang Ayi said. But Rainey didn't budge.

"*Bu tinghua*," she said, putting down the spoon. *Doesn't listen. Tinghua* was the listen-and-obey command Teacher Chen had issued on Rainey's first day of school. That day Rainey had failed the *tinghua* test, and he'd flunked this one, too.

Tang Ayi headed to our front door and started putting on her shoes. I followed her to the door, incredulous.

"Excuse me, you're leaving?" I asked.

"I'd heard foreign children *bu tinghua*." And with that, she shut the door behind her. Ayis might prefer American employers, but a US passport didn't trump a disobedient child.

Wu Ayi came the next day for her trial. She revealed that her husband was a construction worker in another city, and her own wages were designated for tutoring costs for their son.

All seemed to go well until afternoon nap, when Rainey kept popping up like an overwound jack-in-the-box in his crib, whimpering and crying. Wu strode over to the crib, jerked Rainey off balance so that his body toppled forward, and pushed him into a lying position. "*Tinghua*! Listen! Lie down! Hold your head still."

She held Rainey's head against the mattress, and I watched in disbelief as his arms began flailing. Within two bounds, I was crib-side.

"Let him go," I said, removing Wu's hand from the back of my son's head. Rainey lifted his head and continued to cry.

"He doesn't listen," Wu said, glancing up at me. I recognized the reproach in her glance. This was my fault!

"I don't think we're a good match," I told her, my fist clenching. "I'd like you to leave."

By now the word had surely gotten out: Rainey was a terror. I called Carol to let her know I'd lost two prospects, with only Hu Ayi remaining. But Wu and Hu must have been in cahoots, because Carol soon delivered the news: Hu Ayi thought Rainey was "very young and our house was too difficult to clean." (Never mind that Hu Ayi had shown up at our interview looking as dirty as a chimney.) Our final prospect would pass.

The subtext was clear, reasons carefully calibrated so I wouldn't lose face. "You're a terrible parent. Your child is a nightmare. As much as I need this job to raise my own child, I can't take care of a foreign kid who doesn't listen."

★ ★ ★

OUR FOREIGN KID has two parents who set about tackling their new lives with purpose and a sense of adventure.

It helped that our marriage was itself forged, in some sense, on an appreciation of differences; one of us was always adjusting to new circumstances. The first time I flew in to meet Rob's parents in Minnesota, I stepped off the plane from New York, onto an airport tram, and immediately noted that I was the shortest person there, and the only one with hair darker than white chocolate. I'd spent most of my life in the melting-pot megacities of Houston, San Francisco, and New York City—I'd never before seen so many tall, blond titans gathered in one place.

See, Rob and I hail from two cultures, as distinct as lutefisk and lychees.

Rob grew up in a Minnesota town with a single stoplight. His predecessors were Swedes, Norwegians, and Germans who'd settled in the Midwest generations ago, and if you'd lined up all the townspeople on Main Street—and thrown in the apparitions of their ancestors for good measure—the people would still be outnumbered by the frogs and fish in the lake out back. Most everyone was on a first-name basis with the druggist and dentist, and the community's anchors were sports and church. A neighbor might recall, game by game, the high school hockey prowess of your uncle, and when someone in your family died, people materialized on your doorstep with casseroles. Rob's parents set curfews and firm rules about school attendance, but otherwise Rob and his brothers spent hours outside playing ball, exploring the woods out front, and swimming or boating the lake behind their house.

It was an idyllic childhood, but it was also insular, and Rob never looked back once he found the opportunity to leave. During his twenties, he hopscotched from one continent to the next, living in Spain, then Australia, and later China during the 1990s as a Peace Corps volunteer. In fact, Rob was one of the first foreigners to live in the town of Zigong in rural China since Mao Zedong's takeover of the country in 1949. His second day there, Rob pulled aside the drapes of his ground-floor apartment to find half a dozen Chinese children, hands grasping the bars outside his window, eager for a peek at the American arrival. Rob mirrored their wonder, turning his toward the place that would be home for the next two years. China's was a rapidly changing society, yet letters still took two months to reach the United States, and email wasn't yet commonly used to communicate with friends and loved ones. Isolation from the outside world granted Rob the focus and the time to learn Mandarin, devour Chinese history books, and befriend his neighbors.

My connection to China came by birthright. I am a direct descendant of the founding emperor of the Ming dynasty, but Mao Zedong's armies cared little about dynastic heritage when they came marching through China in the 1940s. During this time, my mother's and father's families fled China and near-certain persecution, seeking safety and a stability the country couldn't provide. "It was hard," was the only phrase my maternal grandmother would ever utter about her travails, with the emotional reserve characteristic of many Chinese. For decades after that, the routes back into China were shuttered, and no one saw extended family for years.

The uncles, aunts, and cousins who stayed behind in China braved war and Mao's devastating campaigns, including the 1960s-era Cultural Revolution, and only began to prosper as China opened up over the past few decades. As China rose, so did they: Some became prominent executives and politicians. Most notably, my granduncle Zhu Rongji developed Shanghai as mayor during the 1990s and ultimately became one of the most acclaimed premiers of modern China.

Across the Pacific Ocean, my mother and father had emigrated as youngsters to America, where collecting advanced degrees must have been a dating ritual: They met in Michigan, earned Ivy League PhDs together, got married, and settled down in the suburbs of Houston. There, they would raise me and my baby sister almost with a sense of urgency—it was an anxiety born from knowing prosperity in a faraway land, only to be forced out with nothing but their lives, their education, and a few bars of carefully smuggled gold.

My family line started anew in America. I wish I could say that childhood was fun, that my parents tackled their new lives with the adventurer's sense of spirit. Instead, I grew up under the invisible hand of ancestral expectation, which clutched my shoulder with the intensity of a vise grip. Certainly, I never knew war or revolution, and I attended American public schools—which in Texas revolved around the rituals of football and cheerleading—but I came home to my parents' authoritarian Chinese ways.

My mother and father wielded their authority without mercy, plotting mine and my sister's paths to test perfection and advanced degrees. Unlike for Rob's, church and sports weren't a compass for my family; we worshipped at the altar of education, and if I'd tried to apply to the US Peace Corps, I'm sure my father would have asked why I'd choose to live in a developing country when my predecessors sought to escape their borders. ("Why go somewhere you can't drink the tap water?")

Rob and I met in New York City in our late twenties, as graduate students in journalism. By then, we'd traveled six continents between us, hopscotching between jobs and backpack trips around the world, and the wanderlust continued after we got married. We shared five homes in as many years. Rob pursued public radio journalism while I became a newspaper reporter and later a writer.

Moving to China proved a homecoming for each of us in a different way. Rob's time in the Peace Corps in China had inspired his decision to become a journalist. My ancestry in China meant I had the frame of reference of a descendant who'd left for an extended expedition and journeyed home for a visit. Rob and I both had just enough of that change-seeking enzyme in our blood to pack up a baby boy and relocate to a foreign country.

This spirit of adventure and adaptability would prove helpful as we grew into our roles as parents, and it would become critical as we learned what was expected of us—not only from Chinese teachers but from society—as parents of a child in China.

★ ★ ★

SOON ENOUGH, at Rainey's school, I was getting signals that a child's education should be a full-time job for at least one parent. Three months into the school year, I met a Beijing woman who'd abruptly left her job selling equipment during China's economic boom. Brokering industrial machines had made her wealthy, but the job kept her trapped in meetings at unpredictable hours.

"One day, I had to choose between a meeting with my big boss and an activity with my daughter at school," she told me. "They were scheduled at exactly the same time."

In the kerfuffle, she somehow missed both appointments.

"The school proved *less* forgiving than work. So I quit my job," she said.

Soong Qing Ling liked this kind of parent. "We don't pick children—we pick parents," Principal Zhang had told a packed auditorium of bobbing, black-haired heads during school orientation. "Yesterday, I met a parent who purchased *two* flutes. One flute for herself, so she could practice alongside her child. I like that kind of parent."

Principal Zhang also liked parents who read and respond. There's a lot of back-and-forth in a Chinese education: notices, memos, text messages, emails, letters, and permission slips. A series of bulletin boards hung outside Rainey's classroom, and they courted a daily gaggle of parents and grandparents who crowded around, two and three bodies deep. I came to call these five-foot-square monstrosities "Big Board." Its mainstays were the week's schedule, alongside the lunch menu and a scattering of student artwork, say, two dozen peacock drawings, with each tail feather parsed out at exactly the same angle. It was Big Board's directives I began to struggle with, those memos that disseminated all kinds of tasks for parent and child: work to complete at home, books to read, tasks to perform on behalf of administrators. When missives to parents weren't posted, they were sent home with children in a three-ring binder, into which teachers slotted articles and notices. There was also a classroom blog.

Most bothersome of all was the WeChat parent group, which kept me tethered to my phone at all hours of the day and night. WeChat is an instant messaging platform more popular than email as a means of communication in China. Through this mobile app, the teachers could communicate to us their every command. Their instructions were fast and frequent:

> This week we're starting the "I Love My Family" theme. We'll start by painting pictures of the mother. Mothers, please bring a photo of yourself to school. Please reply in the affirmative.

Keeping up with daily instructions like this were part of the job. Sometimes, the teachers' directives were bewildering: "Bring plastic fish to school"; or troubling: "Health check is tomorrow, so tell your child he must endure the finger prick without fear"; or antithetical to my belief about the strength of the human body: "Today the weather is cold so children cannot play outdoor sports."

Other times, teachers deployed parents as free labor: "Tear out all your children's worksheets and affix them to the corresponding lesson page." That day, Rainey marched home seven workbooks, and we set up a mini-factory on our dining-room table, tearing worksheets, reshuffling papers, and stapling for hours.

"Shouldn't this be the school's responsibility? Or a teacher's aide?" Rob asked, picking up a stapler. I shook my head.

Parents are full partners in a Chinese education, and pity the mother who overlooked a message and failed to bring plastic fish to school.

As the weeks passed, the barrage of daily WeChat messages continued, and my anxiety levels continued to creep up. A parent's reply to a teacher's WeChat message was expected to be immediate, if not instantaneous, and keeping up with this daily flow of information was part of my job. Before long, I realized what troubled me most: the other Chinese parents. It all began to feel like a race, as if we were playing endless rounds of musical chairs, and the last parent to respond would have her supports immediately kicked out from underneath her. A teacher would message:

> Do any parents own "tortoise and hare" clothing for a role-playing exercise?

Within seconds, a cacophony of mostly mothers would chime in, and my phone buzzed as each message arrived:

> Yes! I do! Teacher, you work so hard!
> I have tortoise and hare clothing!
> I will go shopping immediately! Teacher, you are too amazing!
> I received the message! We will do as you say!

One overachieving parent wrote that she had not only tortoise and hare clothing but also "frogs, goldfish, *and* tadpole clothing!" Others immediately launched a search for animal costumes (or at least, they announced their diligence on WeChat).

I didn't see the point of the role-play exercise, nor did I care to dress my three-year-old son in reptilian clothing. But I wanted to keep my chair on this particular day, so I broadcast my own enthusiasm:

Teacher, will do right away!

The teachers' own messages came at all hours, even late at night and on weekends, and on some days I counted north of three hundred messages buzzing around the group.

Most Chinese families had an easier time keeping up than I did: It was simple mathematics. China's planned-birth policy of the last few decades—the "one-child policy," colloquially—meant that many of Rainey's classmates have no siblings. These children have one mother, one father, and four grandparents to share parenting duties—that's six adults per child. Living arrangements reflect this top-heavy arrangement in major Chinese cities, and it's not uncommon to see three generations under one roof: kid plus parents, with a side of grandparent or two.

Sometimes this arrangement backfires, cultivating a child so spoiled that pop culture deems them "little emperors." Other times, lots of helping hands transfers a crushing pressure of expectation onto the student at the bottom of this pyramid. I often think of the Chinese proverb *bu kan zhong fu*, conjuring up a donkey or workhorse, piled high and weighed down with goods, that can "no longer carry the burden."

On the positive side, a six-to-one adult-to-child ratio means that Chinese urban children generally have more warm bodies around to share the duties of education. My granduncle Kuangguo, a talkative man with a commanding poise honed from decades of sitting at the heads of banquet tables, takes to grandparenting like a blood sport.

"My most important work is only beginning," he told me, now that he has a grandchild to dote over. Uncle Kuangguo carts stacks of cash for extracurricular activities, lords over homework, arranges gifts for teachers, and attends field trips. His granddaughter actually relocates to his apartment during the school week—she has her own bed there—with the Monday-to-Friday migration made easier by the fact she has only to get in the elevator and press a button. The girl's father, my second cousin, bought an apartment in the same building as Kuangguo just for this purpose.

By contrast, Rainey has only two reliable adults in his life, Rob and me. Rainey's grandparents were an ocean away, so we couldn't tap extended family to crowd-source child care, and I strained under the weight of keeping up with directives written in my second language. I was finding fulfilling work as a writer, and later as a television correspondent, and I wasn't about to abandon that to appease Teacher Chen.

I thought I'd mastered the balancing act as well as the guilt, until Grandparents' Day came around to prove me wrong. Nothing flaunted the delicate balancing act of our short-staffed household like this particular holiday.

"It's a virtue of our Chinese nation to respect the old people. Our grandparents have paid so much love that we should do what we can as they grow old," a teacher's note told us one day. The note instructed us to send one grandparent to school that Thursday to mark a celebration of "filial obedience."

This would be a problem.

"Rainey's grandparents live in the United States," I told Teacher Chen at pickup.

"Well, then Rainey can't participate," Chen replied. "Keep him home from school."

"*Buhaoyisi*—my apologies—I'm sure Rainey would like to participate," I said.

"Then you come with him," Chen said, briskly.

"I work, and I'll be unable to get time off that day," I said.

"Then—keep him home from school."

So, that's how I found myself begging off work for a day to impersonate my own son's grandmother. I walked into Small Class No. 4's classroom two minutes late to find a couple dozen grandparents seated in tiny chairs, singing a song from 1952 about labor.

Working is the most honorable thing
The sunlight is shining and the rooster is singing...

I glanced over at Rainey. He seemed strangely disengaged, and he avoided my eyes.

Flowers wake up and birds are preening...
A little magpie sets up house, and a bee gathers honey.

I looked over at Rainey again, but he seemed intent on ignoring me. Was that because his grandmother for the day in fact gave birth to him? I didn't expect a three-year-old would feel peer pressure, but right then I realized the ways in which Rainey was different from his classmates were sometimes impossible to hide. This was made worse by the Chinese cultural focus on the collective rather than the individual, and I saw that Rainey's psychology was beginning to shift in that direction. He was embarrassed because he stood out.

The children slapped their knees to "Big Rooster, Big Chicken," and then another jolly tune built around the Party's themes of work and labor, and finally Teacher Chen arrived at the point of the holiday: a lesson on filial piety.

"Children!" Teacher Chen exclaimed, clapping twice. "We should tell our grandparents, 'You work too hard!' Your grandparents are older, and as a show of respect, let's massage our elders!"

Massage our elders?

Chairs squeaked as children rose and made beelines for the backs of their respective grandparents. Seated in a miniature chair, I was eye level with Rainey as he approached.

"Hi, Rainey!" I said, smoothing back his hair.

"*Tai xinkule*—you work too hard!" Teacher Chen commanded. "Say it, children!"

The children chanted. "*Tai xinkule*—you work too hard!" they chirped, gazing up at their elders. Rainey managed to mumble something in my direction.

"Now massage your elders!" Chen proclaimed.

Rainey stepped around my chair and placed his tiny hands on my back. The other children did the same, and I realized the students had rehearsed this moment in class. Rainey tapped halfheartedly, as lightly as you might touch a wall wet with paint, and finally it was over. Twenty-seven children had massaged the backs of twenty-seven Chinese ancestors, while the twenty-eighth—my little boy—had patted the backside of his American mother.

Later that evening at home, Rainey moaned, "I don't have grandparents."

Of course he has grandparents—four of them—but they lived on the other side of the planet. Not one was interested in moving in with us.

As Chinese New Year approached, the most important holiday of the Chinese calendar, I got wind that other parents were planning elaborate gifts. I figured it was time to cultivate a little *guanxi*, or rapport with Rainey's teachers. Rainey didn't have grandparents around to share the duties of his education, but he had one mother very eager to step into the big shoes required of a Chinese parent. Buying gifts was a task I could manage.

I'd found myself at a Coach outlet in Houston during a midyear trip home to see my parents, Rainey in tow. It had been a slow day at the mall, and when the Coach greeter discovered what I was shopping for, she began trailing me with enthusiasm.

"Oh!" she'd said, perking up. "You are shopping for gifts? Gifts for Chinese women?"

"Yes. I need two . . ."

"Yes! For Chinese!" she exclaimed. "The Chinese come here in droves and buy thousands of dollars' worth of purses. They are very good customers!"

Import taxes make foreign brands double and triple the price inside Chinese borders, so the Chinese consumer with means had begun flying to Europe, the United States, Hong Kong, and Korea to shop. The Houston saleswoman spotted an over-the-shoulder satchel with bright blue Cs on it and tapped the letters with her forefinger. Out tumbled her diagnosis of the buying habits of well-to-do Chinese: "The Chinese like anything with the 'C' on it. They love the metal insignias with the Coach wagon and horses on it, because it is the classic logo. They love patent leather and bright shiny plastic colors. They like to have a zipper closure so nothing is open and vulnerable. And they like an over-the-shoulder strap," she concluded, slipping the bag over her shoulder. When she spoke, her words began to come out faster and faster as she fingered the corresponding part of the bag—logo, leather, zipper, strap.

I dropped $500 on handbags, each with the four must-have features, hoping for a little attention for my son. Rainey was confused. During the drive back from the mall, he'd sung out, "Who are those purses for?"

"For your Teachers Chen and Cai."

Puzzlement crept into Rainey's voice. "They already have purses," he said.

"It's a gift," I said.

"So, they will each have two purses?"

"Yes," I answered, focusing intently on the road past the steering wheel. Rainey had already sussed out the absurdity of the situation.

In modern China, gift-giving is an important ritual inside any valued relationship. All types of implications are suggested based on what the gift is, how expensive it is, and the manner in which it is presented, with machinations sometimes bordering on the comical. Most Chinese I know find this ritual exhausting, and the act itself is so complicated that researchers have devoted entire dissertations and studies to teasing out its formalities and insinuations. Chinese culture is hierarchical, and "respectful verbal and nonverbal behaviors" must be nailed when presenting a gift to someone of "high power status," wrote academic Hairong Feng.

In other words, it was never as simple as handing over a tissue-wrapped box. Teachers Chen and Cai were of "high power status" in our relationship, and my son spent eight hours a day in their care. In academic terms, I needed to show my appreciation for the teachers and also generate a feeling of reciprocity.

In short, the exchange had to happen smoothly.

How would I accomplish this? I envisioned excited teachers accepting my bounty with open arms and broad smiles, but I wasn't sure how to arrive at that vision. A Chinese friend said I should be discreet, so I thought about slipping into the classroom at pickup time and leaving wrapped boxes in their cupboards, but I wasn't sure what to write in a note. Another friend suggested I casually hand each teacher a slip of paper with an address. "They'll know to arrive at the address and expect a box to be handed to them." That was the Chinese way, she told me. But that seemed too Mafia, too Triad.

Finally, I decided to play messenger. I placed each purse in a box emblazoned with the Coach logo and tossed them in a plastic bag. When I arrived at the classroom door, I addressed Teacher Cai in earshot of a Shanghainese man who was collecting his granddaughter's things. But, as I approached Teacher Cai, the bag broke and the logo-imprinted boxes spilled into plain view.

"We have gifts for you from America," I said, uncertainly, head bowed as I approached Cai holding a box I'd retrieved from the floor. The head bow, I thought, was a "respectful nonverbal behavior" indicating her status.

Teacher Cai's reaction was emphatic and immediate.

"*Bu yong, bu yong*—no need, no need," she exclaimed, hands raised with palms out as she backed away from me. She glanced at the grandfather, who retreated quickly from the scene. After Teacher Cai moved a safe distance away, she pivoted and walked rapidly in the opposite direction, leaving me at the door.

"*Bu yong, bu yong*," she repeated, over her shoulder.

I had received both verbal and nonverbal rejections of my gift, and I stood there like an idiot. Thinking through what I knew of Chinese gift-giving practices, I knew this type of reaction in an asymmetrical power relationship would call for more "reoffer-decline exchanges."

In other words, the recipient is *supposed* to decline a gift the first few times, and the giver is expected to try again. I'd watched this custom unfold when my parents met with a Chinese friend for dinner. Even if the occasion clearly called for my father to pick up the check, his guest would decline the first, second, and third times my father insisted. This usually resulted in a verbal—if not physical—tussle over the check.

"No need!" the guest would say, putting up a hand.

"You treated last time!" my father would say.

"Really, it's my pleasure to spend time with you, your presence is your gift to me," the guest would urge.

"You must let me pay . . ." my father countered.

The Chinese server would stand by, waiting for the circus to right its top. Once I watched a guest chase my father around a restaurant table three times in pursuit of the bill.

I had no such tolerance for melodrama, and Teacher Cai's reaction mortified me. I didn't care to participate in any kind of "reoffer-decline exchange," and I biked home with heart sunken into the toes of my shoes, the damned Coach purses bouncing in the basket of my bicycle. There was a distinct drive underlying the existence of the Chinese family—whether it was the migrant ayis working big-city jobs to send money home, urban dads sitting in math cram school waiting rooms, or the Soong Qing Ling moms scrambling to post on WeChat—and I was trying to play my part as the right kind of Chinese mother.

Once home, I shoved the boxes deep into the closet, behind our coats.

Clearly, I still had much to learn.

3

OBEY THE TEACHER

Sit DOWN or your mommies won't come pick you up after school today!

—TEACHER WANG

The Chinese traditionally have little respect for animals, as they historically served one of two purposes: They are for eating or for pulling equipment.

One Saturday that fall, Rob, Rainey, and I spent an afternoon at the Shanghai Zoo, where I saw that crowds were rowdy and disrespectful toward the animals, while zookeepers looked the other way. We visited a ferret housed in a glass box that appeared to be designed for a snake. As Rob recounted the time, in rural Sichuan, that zookeepers let random visitors throw live chickens to tigers at lunchtime, I watched a man hurl a glass bottle at an orangutan while his friends cackled with laughter. Inside Shanghai Zoo's primate building, Rob, Rainey, and I gazed upon its resident gorilla, trapped inside a concrete cell the size of my living room, with a low ceiling, no foliage, and a wall of windows behind which human visitors stood to gape.

As we prepared for dinner that night, Rainey began leaping around our living room on long arms. "Rainey Gorilla, Rainey Gorilla," he announced. "Gorilla sad."

"Why, Rainey?" I asked him. "Why Gorilla sad?"

"Gorilla all alone. Mommy, Daddy away in jungle," he said. "Gorilla at school."

My heart jumped. "Gorilla *at school*?" I asked, thinking of the gorilla's bare, solitary cell. "Does Gorilla have friends?"

Rainey didn't answer.

The following week, the complaining began in earnest. "I hate school. I haaaaate school," Rainey would whine, a low, constant, dull moan that got right into that register where it seemed to bounce off the eardrum and echo for a few seconds before it finally dissipated. Which was exactly the moment Rainey would start all over again.

"I haaaaate school."

I never believed that an education should be 100 percent fun, but I didn't remember hating school to the point that I whined over my breakfast oats. "Why, Rainey?" I asked him. "Why don't you like school?"

"Every day school," Rainey said. "Next day school. After that, school. Always school, school, school."

I thought of the times I'd lower my head into my pillow at night, relieved to shutter my eyes on a vexing day. It was natural to experience an adjustment period to anything, especially to living in a foreign country and attending school with Chinese classmates. Was Rainey simply going through his own adjustment process?

I began listening to his whining, hoping for clues. One day, he got specific.

"I was sitting very well but the teacher still got mad at me," Rainey said. "I don't know why." Another time, he blurted, "The teachers are always loud. I don't want to go to school. They yell. It makes my heart hurt."

They yell? Loud? I thought. How loud, and what did that mean, exactly?

"How do they yell?" I'd ask Rainey.

"I don't want to talk about it," he'd always answer. The Chinese generally have the loudest voices of anyone I've ever met (my own father's voice could, it seemed, project from one side of the Grand Canyon to the other). Was it possible that Rainey's three-year-old ears were simply sensitive? I'd come to terms with the force-fed eggs, and some days he seemed to like school, especially the times he came home with a shiny star sticker on his forehead, which he paraded around proudly. Even so, these new clues were disturbing. What was going on inside Teacher Chen's classroom? Was there reason for concern?

My only real glimpse inside Rainey's day came via the three-ring yellow binder Teacher Chen sent home each week officially entitled the "Child Development Book." It showed snapshots of the children in various parts of the classroom: posing as chefs near the play kitchen, lined up in rows with mouths open for singing class, or standing single file with playground equipment in the background.

One photo showed Rainey, his face unmistakably sullen, propped in the lap of an older boy, whose arms encircled Rainey's waist. The caption alluded to a mentor program: "Sisters and Brothers of the Big Class Take Care of Small Class. The Small Class no longer miss their mothers and fathers during the daytime—they are becoming independent and brave." I'd never seen Rainey looking more miserable.

The binder also contained instruction sheets for parents. One advised how parents could work to improve a child's "bad habit" of "poor concentration." Another advised that "children should be taught to greet fathers after work with a cup of water and slippers." A tip sheet apprised women of their maternal duties: "The mother's major task in autumn is to protect kids from dryness-heat. Make brown rice or pear porridge to prevent dry throats, dry lips, nosebleeds, and dry skin." Apparently, the ideal Chinese father was away toiling at work, mothers were installed in the kitchen, and children existed to serve the paternal head of household.

Other than what I read in the Child Development Book, I knew very little about the teachers' classroom style or educational philosophies. There were hundreds of WeChat messages but little *meaningful* communication from them. I gathered that some other Soong Qing Ling parents were as frustrated as I was, particularly the few foreign parents whose children were sprinkled throughout the five classes of Small Class grade. We foreign parents—from America, Australia, France, and Japan—began quizzing our children and sharing the results.

One woman told me her son had simply stated, "We sit there."

"Sit? Do you also read books? Sing songs?" she'd asked.

"No, we just sit there."

"And do what?"

"We sit there and do nothing," the little boy had said.

Another mom told me her son said they go outside and balance sandbags on their heads, and that "he seems happy with this activity."

I'd begun throwing questions at Rainey during dinnertime: "What do the teachers say to you? What does the school ayi say to you? What do you do during school?"

Rainey would not respond. But as the days passed, my son began to use my desperation as leverage:

"I'll tell you about school if you don't make me go to bed tonight."

"Buy me some gummy bears and I'll tell you about school."

And finally, "I'll tell you about school—if you let me stay home from school."

I began creating excuses to visit Soong Qing Ling during the school day, showing up just before lunchtime to drop off an extra shirt for Rainey (making sure children were toasty warm was of supreme importance to Chinese parents), or paying school fees in person rather than by bank transfer. (Public elementary and middle schools are typically free under the government's compulsory education plan for grades 1 through 9, but kindergartens—which take children as young as three—aren't yet compulsory and charge tuition.) Each time I oh-so-casually sauntered past the door of Small Class No. 4, looked around to ensure no adult was nearby, and tiptoed over to place an ear at the door. I could never hear anything. I felt like a crazy person.

Soong Qing Ling was impenetrable. One day I tracked down

Principal Zhang after school, hoping to get permission to stand in back of a classroom—any classroom—for a few hours. At pickup time, Zhang was standing at the edge of the green lawn, watching families stream out of the school gates, and I stepped up carefully, making sure I didn't block her field of vision.

"Hello, Zhang Yuanzhang, I'm Rainey's mother. Thanks for allowing us into the school," I told her. "We are very excited to be here."

"*Hao, hao,*" she said. Good, good. I told her I thought the playground equipment was impressive and also made some comments about the weather. Then I took a deep breath and plunged in.

"I'm wondering, would I be able to observe a class someday?"

"We don't allow people to watch classes," she said, glancing at my face and then back to the families leaving the school grounds. I launched the flattery technique.

"Well, I hope to learn about China's educational style," I said. "In the West, we think the Chinese education system is so impressive!"

"We don't allow observations by non-teachers," she said, looking over my shoulder and then carving a hasty exit. "Excuse me."

I became determined to see the inside of a Chinese classroom. I decided to try the next best thing in my reporting mission: another Shanghai kindergarten. Exactly what kind of environment was I throwing my son into? Rainey mentioned sitting—why was this so important? Is Chinese education about conformity and concentration? Were the teachers' methods as harsh as I feared?

I asked some Chinese acquaintances to pull on their *guanxi*, and I finally received an invitation to sit in on a classroom at Renhe, or Harmony, Kindergarten. It wouldn't come until about six months from now, at the start of the first week of school the following academic year, but I knew it would be an important opportunity. Shanghai had more than fourteen hundred public kindergartens, all governed by central Ministry of Education guidelines, with oversight by the local education bureau, and I was certain I'd learn something about Rainey's environment through a peek inside this classroom.

My domain for observation would be a class of the youngest children, most of whom were three, like Rainey. Because it was their first time away from home, here was a unique chance to observe teachers managing behavior in the most challenging of circumstances.

Access came as a favor for the friend who'd asked. My friend made clear what the "unspoken" terms were.

"A gift for the teacher would be expected."

"Okay—what would be an appropriate value?"

"Say, one thousand kuai?"* That was $160, and about a quarter of a teacher's monthly salary. In China, when you needed something to happen, tactical prowess was never as effective as a main course of *guanxi* with a side of gift.

"Would a Coach purse work?" I countered. "I have a few handy."

"Yes. Coach is good."

* * *

I ARRIVED AT Harmony Kindergarten at 8:32 a.m. and knocked on the door of Small Class No. 1, the sounds of wailing wafting through the locked door.

Teacher Li answered, a slender thirtysomething with a pixie cut and eyebrows drawn in thick, black pencil. "Today's going to be *luan*—chaotic," Li said as she ushered me inside, locking the door behind us with a click.

It was the first day of school at Harmony Kindergarten, and I found twenty-eight tiny, wandering children in various stages of distress. Most were crying, and some were uttering variations of the refrain: *Mama! Mama! I want to go home!* They'd spent their first three years of life in the coddled comfort of home, with parents and grandparents lording nearby. That had ended abruptly with their enrollment in kindergarten and the commencement of the long, narrow road that is a formal Chinese education.

* Kuai and yuan are different names of the same basic unit of Chinese currency.

Fittingly, that road started with an order. Master Teacher Wang was in charge of today's lesson, a fearsome woman with long black hair that seemed to pull down with it all of her facial features as the strands fell toward the floor. Wang had sharp, focused eyes, a downturned mouth, a chin that seemed to point and jut, and a staccato voice that made you jump, since she liked to use sound as a weapon of surprise.

"Sit . . . sit . . . sit down . . . sit *DOWN*! *Sit DOWN* or your mommies won't come pick you up after school today!" thundered Master Wang.

The children were meandering about, weaving between tables and chairs, as if some giant had taken a handful of roly-poly Weeble figures and tossed them into a shoe box. Some had come to rest, while others were wobbling around the room in search of something familiar. All looked confused. *Mama! Mama! I want to go home!*

Teachers Wang and Li were the masters of this classroom, and their goal this first morning was twenty-eight little behinds planted in twenty-eight tiny chairs. The wooden chairs were arranged in U-shaped formation to face the front of the classroom, which measured the size of a two-car garage. The room was packed with the trappings of a Chinese education: bunk beds stacked high against three walls, a dark chalkboard, two porcelain chamber pots already filled a few inches deep with yellow pee. (Rainey's school was a touch more modern, although some bathrooms there contained squat toilets.)

"Sit down!" Wang and Li marched around the room, eyeing the whimpering children. With nearly every step they would encounter a small child, and with a swift motion they'd grasp an upper arm and maneuver a tiny body into the nearest chair. Both teachers moved deliberately, but Wang in particular had a way of sucking all the air from the room as she traveled, like a robotic vacuum cleaner that lacked an off switch.

"Sit DOWN!" Master Wang said. "Sit DOWN or your mother won't come get you today. Sit DOWN or your grandmother won't come get you today! Sit DOWN or I won't let you go home after nap time."

The children cried harder. *Mama! Mama! I want to go home!*

The din was phenomenal. The teachers were screeching over the noise. Wang was tall and thin, all angles in her face. She had a sharp voice that could turn from sugary sweet (*Hello, Miss Chu*) to machete sharp (*Stand still!*) in quick succession. Her movements were also abrupt: pointing toward a vacant seat; rapping on a table three times; squatting suddenly to lift a child by the armpits and put her in place. I thought about Rainey's first day, and wondered how Teacher Chen had conducted herself with my child.

Once most bodies were in position, the refinement of sitting began.

"Little hands on your legs! Backs straight! Little feet side by side on the floor!" Wang backed up verbal orders with physical action, a potent combination. She would kick a misplaced foot into place, grab flailing hands and crush them flat against thighs, nudge backs straight with a knock against the shoulder blades.

After watching for about five minutes, I spotted an empty chair and sat down, feet side by side.

I could already pick out the troublemakers. One boy simply couldn't still his body upon command. He was big for his age, with a head nearly the size of a pumpkin and a broad body to match; he wandered about the classroom aimlessly. In America, physical size might draw predictions of athletic achievement—"he's a future linebacker"—but in China it only made you easier to spot when misbehaving.

"Wang Wu Ze, sit down! What is wrong with you? Come and sit down in this chair right now!"

Little Pumpkin sat for a few moments, only to pop back up.

Master Wang put him back in place with a push of a shoulder. Pop up, push down, pop up, push down. This whack-a-mole game would continue all day long.

Children who didn't sit were admonished. A little girl who wandered over to the water cooler was told: "It's not time for drinking water yet. Sit down." Another girl was enticed by a play kitchen and two pieces of plastic fruit in the corner. Teacher Li spotted her and in two flashes bounded over, lifted her by the armpits to standing, and placed her back in her seat. Not a word was said.

By ten a.m. I needed to go to the restroom but I was afraid to cause a disturbance by standing. The teachers hadn't told the children they could not get water without asking, play with toys in the corner, or speak without being spoken to. Yet when children overstepped boundaries they didn't know existed, they were corralled back into place. It was learning by trial and error, and I imagined that the one thing that would quickly become clear to a child was that sitting quietly with feet and hands in place was the safest thing to do. So when Pan Li Bao approached me, I was fearful. She was a chubby girl with two spindly black pigtails, and she had something to say.

"My mom still hasn't come," Little Pigtails said to me, eyes imploring. I glanced at Wang. She hadn't seen us yet.

"You cannot come back here," I said to Little Pigtails as quietly and forcefully as I could.

"My mom still hasn't come. My mom went to work," she said, grasping my forearm.

"Your mom will come in the afternoon," I whispered fervently from my chair. "Now sit down!" Master Wang spotted us at that moment, and in a bound she scooped up Little Pigtails, depositing her back in the U-shaped formation. She'd stepped over my legs to squeeze past, and I thought I detected a glance of disapproval.

In another minute, Little Pigtails was back. This time she put a furry brown stuffed animal in my lap, a moose wearing a T-shirt with a green M&M character on it.

Pigtails pointed at the M&M on the T-shirt, focusing on its oversize white eyes with googly black pupils. "*Wo hai pa*—I'm scared," she said. I glanced around the classroom. There were characters with buggy eyes everywhere. Little Pigtails would have a hard time this year.

"There is nothing to be scared of," I told her. "This is a cartoon character. He is friendly." But she wasn't listening.

"I'm scared, I'm scared," she said.

"Be quiet. Sit down," I said, handing the moose back to her.

I'm scared, too, I thought. I'm scared Master Wang is going to boot me from this classroom for creating a fracas. Little Pigtails chose that moment to climb into my lap.

"Take it away, take it away," she said, trying to put the moose in my hands. Oh what the hell, I thought. I grabbed the offending moose and placed it under my chair. I glanced over at Master Wang, who was working on Little Pumpkin. Pop up, push down. Pop up, push down. Soon Wang's razor-sharp sense for bodies out of place would be drawn to my corner of the classroom.

"Go sit down," I whispered to Pigtails. She didn't move. I was desperate and grasping at whatever I could to make her obey. "Go sit down or your mom won't come get you today!" I said firmly. The moment the words came out, I felt ashamed.

"Mama!" wailed Little Pigtails. I wanted to comfort her, but she was out of place and soon enough Master Wang's ire would focus in on me. The urgency hit me, rising up from my belly into my throat, and I yelped at the little girl: "Go sit down!" I twisted out of her grasp and gently nudged her off my lap. I pointed toward her chair.

I was no better than the teachers. But order had been restored.

* * *

BY DAY TWO, it was clear to me that Master Wang and Teacher Li had an implicit good cop–bad cop arrangement. Li would instruct class while Wang would hover over children, eyeing hands and feet to ensure proper placement.

"They're pretty used to it already," Li had told me that morning, with a nod of satisfaction toward the children sitting in a U-shaped formation, while Master Wang marched nearby.

On day three, the teachers began to explain to the children what was expected of them. It was the clearest directive yet on the rules of the classroom. The instruction came in the form of a song:

I am a good baby
Little hands always in place
Little feet refined
Little ears listening well
Little eyes looking at the teacher
Before I speak, always raising my hand

The teachers had the children sing along, urging them to keep time with their hands. They reinforced the message with candy. A visiting teacher, Teacher Tang, came in with a plastic vial of Skittles.

"Isn't it fragrant?" said Tang, marching around the U, just as Wang had the day before. She shook the vial as she walked, and the candy made a jolly, tinkling sound.

"Take a sniff," she said, pausing before each child. She tipped the container carefully so each of them could get a whiff and a glimpse of the colorful pellets inside.

"Aren't they colorful?" she asked the group.

"Yes, Teacher!"

"Who's sitting well? Whoever's sitting well will get a piece of candy," Tang said.

Several children piped in. "I am, I am! I am, Teacher!"

Teacher Tang made a show of examining placement of hands, feet, knees, before nodding approval and doling out candy, piece by piece.

The next day, the same exercise was performed with red star stickers. "You can't go home unless you get a red star," Master Wang stated clearly while she walked the U, appraising each child. As she pressed star stickers onto foreheads, she made an example of every recipient.

"This student didn't waste any rice today, so he gets a star."

"This student fell asleep quickly at nap time, so she gets a star."

"This student sat nicely today, so he gets a star." I now better understood the significance of Rainey's early days in school, when he came home with a star in the middle of his forehead. I wondered what he'd done—or not done—on the days his forehead was clean. I also realized

Rainey's developing habit of bartering and negotiating—"I'll tell you if you let me watch Thomas the Train"—might spring from the action-reward loop Chinese teachers deployed.

Clear rules began to emerge for Pumpkin and his classmates: Don't break snack-time biscuits in two. Water can be drunk only during water breaks. No talking while lining up. No talking during lunch. Open your mouths wide "like a lion" to make room for food. Further instruction was given in song:

When the teacher is talking, you cannot talk.
When the teacher is talking, you cannot get your toy and play.
When the teacher is talking, you cannot wander around.

Bathroom trips happened as a class, twice in the morning and twice in the afternoon, with the children forming a single-file line and walking slowly down the hall along double yellow lines. This was what the teachers called "train" formation, with hands placed on the hips of the child in front of you. Children who needed to pee outside of bathroom time could use the chamber pot in the corner of the classroom. At the end of the day you could lift the red plastic lid to see a gallon's worth of accumulated pee, and possibly even a few floating brown logs, always a source of fascination for the children.

Lunch was eaten in the hallway—there was no space for meals in the classroom—at tiny tables pushed up against the walls. A meal might be a quail egg over steamed broccoli, chicken and rice, or a section of Chinese sausage laid on top of fried noodles. Children were urged to finish their meals, and the consequences were clear: "Do you want to get up from the table? Then eat your rice. If you don't eat the egg, your mom won't come to get you today."

Is that how they got Rainey to eat eggs? I wondered.

The teachers were not without kindness; at times—usually when the children were sitting—Master Wang would flash smiles at her tiny charges. Another time, a little girl had been bitten by another classmate, and Wang held her for a few minutes, stroking her hair. Meanwhile, the biter had been positioned in a chair at the front of the class, made to face his twenty-seven classmates, who stared past him for a half hour while a TV blared behind his head. It was a classic shaming ritual, and indeed the boy didn't bite or hit again that day. Little Pigtails was still having trouble with her M&M moose and brought it over to me several times. "I'm scared. Take it away, take it away."

One time Wang came over. "What's the problem?" she asked Little Pigtails. The child's voice failed her, and she began to cry. She could only point at the moose, her sobs increasing in intensity. After some time, Wang adjudicated the matter: "No playing with your toy during class. I'll put the moose here so it can wait for you to finish class. Now sit down!" Wang placed the moose on a shelf six inches from Little Pigtails's chair, M&M's google eyes focused right on her face. Pigtails sat there, staring back, still in tears. When Wang wasn't looking, I kicked the moose under a chair.

By now Little Pumpkin's name was imprinted in my memory, as the teachers had screamed it so many times. Wang Wu Ze, sit down! Wang Wu Ze, put your two feet side by side! Wang Wu Ze, what is wrong with you? Do you want your mommy to come get you today?

Little Pumpkin was terrible at flying under the radar. First of all, he was a head taller than his classmates and full of energy. I sensed this was the most troublesome combination for a Chinese schoolboy—large and lively. Besides that, I'd seen him four days in a row and he was always wearing a brightly colored shirt. He lacked good camouflage. One time he was particularly offensive to the teachers—he had wandered from his chair toward a few toys in the corner during a lesson while Teacher Wang was talking—and she really lost her temper.

"Wang Wu Ze, you don't get a chair. YOU WILL STAND!" In three leaps she was over by his side and swatted his chair away. It fell over, clattered against the floor a few times, then lay still. All the children fell silent, watching, and I also froze in my chair at the back of the room. I was keenly aware that an acquaintance had secured my way into the classroom by presenting me as an observer, and although I was disturbed by what I was seeing, I didn't feel it was my place to intervene in any way. I was gradually becoming captive to this situation.

Little Pumpkin looked at the toppled chair and tears came to his eyes. Suddenly all he wanted was that chair.

"I want to sit, I want to sit." He grasped for Master Wang's arms, seeking comfort, but she flung them out of reach. Little Pumpkin then reached for her hips, attempting an awkward hug, but she stepped away.

"*Bu bao*—I won't hold you," she said to the top of his head. "Do you want a chair? Do you want a chair now?"

"Yes, yes, I want a chair."

"Then you sit in it," Wang said. "If you don't sit in it, I won't give it to you. And your mom won't come get you after school."

⋆ ⋆ ⋆

THE CHINESE ALWAYS have an eye toward efficiency, and sitting accomplishes many goals at once. It imprints the relationship between teacher and student in a physical way, with the master standing tall and the subject relegated to a lower elevation. Sitting is also a convenient way to maintain order in a classroom rammed full of little bodies.

In America, many early educators favor circle time, where children and teachers sit side by side in a giant loop. "Circle up," they'd say, many times a day. Teachers and children gaze upon each other at eye level.

The Chinese I spoke with consider this arrangement extremely odd.

"The children stand up and come in and out of the circle whenever they wish," said Little Pumpkin's Teacher Li, who once stood witness to this phenomenon. "We don't have that luxury in China. You can't just get up and get water when you're in class. There are a lot of kids, and they need to be sitting. You can't just do whatever you want. There have to be *yaoqiu*—standards."

But should three-year-olds be expected to achieve sitting? Aren't the authoritarian methods required to get there a bit harsh? I dare not ask Little Pumpkin's teachers, so I sought out the expert Guo Li Ping.

Guo Li Ping educates preschool teachers at one of China's top universities for early childhood education. His faculty profile states that he specializes in the cognitive development of children, and he has published research reports evaluating the quality of early education in China. I ask him to coffee, my head brimming with questions about disciplinary style.

"The difference between the American and Chinese styles of education has to do with God," he says.

"God?" I ask Guo, who sits across from me in a Shanghai café, sipping a latte. Westerners have the Church and the authority of God, while the Chinese have their teachers, he explained. In the United States, many children attend church from a young age, so the average kid will learn from Sunday sermons when to speak, when to sit, when to pray, and when to break for a meal, he told me. "Internalization of the rules starts from a very young age." Chinese children must have rules of behavior impressed upon them externally.

"And that's where the teachers come in," Guo says, emphasizing his declaration with a lift of his latte. "The Chinese have no religion, and so there is no one else to teach the rules of behavior. Teachers are the ultimate authority."

I wasn't sure about the link between church and schoolhouse behavior, but I found it fascinating that a Chinese researcher accepted such an idea as fact.

"Why do teachers yell in class here?" I ask.

Here, Guo began to make sense. "Speaking loudly is tradition in China," he said. "China is an agricultural society, and in rural areas speaking loudly makes people feel happy and lively." I moved on to threat making by teachers as a classroom technique.

"Yes, this is not what we want," Guo says, shaking his head. "Nowadays we try to teach that threatening children cannot be included in our education system. But in practice, there will be problems. In our traditional culture, teachers are in a higher class than students, which influences the way they treat them. In America, teachers respect children as individuals, but in China the individual is stressed less than society."

In other words, Confucius and the ideal of social harmony still have an effect on classroom dynamics.

I knew from just a few days of observing at Sinan Kindergarten that such harsh methods might be critical to keeping order in crowded classrooms, especially when students outnumber teachers by large ratios. Guo Li Ping confirmed that "with more than fifty children in a classroom, it's simply impossible to let children step out of line. Teachers can only criticize and control. Criticize and control."

Indeed, the numbers for Chinese classroom sizes are legendary. Forty. Sixty. Seventy-five. In the countryside it's not unheard of to cram more than one hundred students into a single room.

Professor Guo acknowledged this problem. He'd spent a year as a visiting professor at Columbia University and visited schools around the United States. His time abroad had impressed him with one thing: Westerners have the luxury of roominess. "If Chinese children could have as much space as children in the West, they would feel amazing," he said.

My Chinese friend Amanda offered a more sinister explanation for a teacher's harsh methods. A student at a top Shanghai high school, she was preparing to apply for colleges in America. We'd been introduced by a mutual friend.

Why, I asked her, are conformity and obeisance so important?

Amanda was fresh from a yearlong exchange at a US high school, which gave her a better understanding of both Chinese and Western education. She'd been reading Nietzsche's *Human, All Too Human*, and a passage leapt out at her. "I think this accurately describes the situation and principle of preschools now in China," she told me:

> The environment in which they are raised tries to make every human being unfree by always keeping the smallest number of possibilities in front of them. . . . We describe a child as having a good character when its narrow adherence to what already exists becomes visible, the child testifies to its awakening sense of community; on the basis of this sense of community, it will later become useful to its state or its class.

★ ★ ★

AT HARMONY KINDERGARTEN at the end of the week, the children were sufficiently subdued that Wang and Li attempted an art class. It would be an exercise in drawing rain.

"Rain falls from the sky to the ground and comes in little dots," Teacher Li said, demonstrating on a piece of paper tacked to a corkboard. She methodically populated a blank white paper with dots that fell from top to bottom, and then filled the page from left to right. The children watched.

Li and Wang immediately fell into good cop–bad cop routine again. Master Wang placed a marker and sheet of paper in front of each child.

"Let's draw the rain," Li said. "Begin!"

"Marker moves from top to bottom," Wang said.

In Li's classroom, rain does not blow sideways, nor does it hurtle to the earth in sheets. There are no hurricanes or monsoons. There is no figurative rain, such as raining tears, or raining frogs, or raining cats and dogs. There is no purple rain. In this classroom, rain is comprised only of tear-shaped droplets of water that fall from sky to ground.

"This is drawn very well," Li said, picking up a paper and showing it to the class. "This is very much like rain," she said of another artwork. "This rain falls nicely from top to bottom," she effused, remarking on another student's work.

I glanced over at Little Pumpkin. He was planted in his chair, but his paper showed no such order. Thick purple streaks crisscrossed the page, right to left, left to right, sideways, diagonally, randomly, obliquely. In one corner he'd simply pressed the marker down and moved it in erratic circles. His world looked like a lava lamp gone psychedelic, with blobs of purple floating across the page. I was afraid for him.

Later, on their way back from bathroom break, the teachers stopped the children from coming into the classroom. They stood single file between the double yellow lines in the hallway, as I hung back near the end of the line.

"Who's standing nicely?" Master Wang asked. Silence.

"If you're standing nicely you can go inside and get some water." Wang and Li walked up and down the line, observing placement of hands and feet, straightness in the back. For about three minutes, they walked up and down the line, hovering over their little soldiers.

Try as I might to figure it out, it wasn't apparent to me what criteria the teachers were using to dismiss children. Nearly all of them were standing nicely with hands by their sides, but the teacher would wait, observe, and then tap a few, wrapping a hand behind a head and giving a firm push. Dismissed: Go get some water. Then they would continue their stroll along the line. Before long, I understood that arbitrariness was the point and identified the exercise for what it was—a demonstration of who was in charge.

Even so, I observed that the children found little ways to express themselves. One little boy was making the shape of a gun with his hand, shooting imaginary critters, feet all the while planted firmly on the yellow lines. A girl was making small flapping hand movements from the wrist, emitting tiny birdlike noises, while keeping her hands by her sides. One boy had a hand down the front of his pants, fumbling around.

I looked at this little group and imagined the 1.4 billion people of China. These children didn't seem unhappy, but they also weren't bounding with the joy, openness, and curiosity that I'd hoped life would be for my son. It seemed to me that there was just a quiet acceptance of fate, of the system, the terms of which only their teachers fully understood and controlled. It was clear that those who grossly overstepped the bounds would be severely punished. Yet—the little birdlike movements, the fumbling in the pants, the shooting gun—they gave me hope that perhaps these children could find ways to express a bit of individuality and push against boundaries, without drawing undue attention to themselves. Or was I grasping for something reassuring?

One by one, the children were tapped to go inside for a water break.

"Wang Wu Ze, do you want water?"

"Yes, Teacher, I want water," said Little Pumpkin.

"Well, you don't get water. You will stand here," Master Wang said, shooting him a glance. She entered the classroom and closed the door behind her.

Little Pumpkin began to cry. His wailing reached depths I'd not yet heard that week, as he stood on the double yellow lines in a cold hallway.

"Teacher, I will sit, I will sit," he cried. "Teacher, let me inside the classroom!" But she was gone.

What to make of a boy like Little Pumpkin? By now he must have known there would be consequences to his actions, yet he was still unable to comply with the rules. In America, such a boy—unbridled energy, passionate purple marker technique—might be labeled as suffering from attention deficit disorder, as having a creative temperament, or perhaps both. He might be called out as a leader: someone who challenges authority, walks to the beat of his own drum, has a way of standing a head above the crowd. But this is China. Would a child like him eventually learn to run within the bounds of the system? Would his individuality continue to make him a target for shame and ridicule? For now, at Harmony Kindergarten, the answer was clear: He would be left alone outside a closed door while his comrades sipped water from metal cups on the other side.

On the last day of my visit, I found the teachers in the classroom when their students were eating outside in the hallway. I expressed thanks for their time and handed over a Coach tote bag and a wallet, which I'd dug out from the back of my closet. Wang and Li nodded and, without a word, stashed the goods in a cabinet. At Soong Qing Ling, my gifting attempt had been an utter embarrassment, but today Coach was my friend.

I left the classroom on this final day and ambled down the hallway, toeing the double yellow line. I passed three classrooms and peeked inside each one to find children seated in chairs arranged in a U-shape around a teacher. When I reached the building exit, I turned around for one last glance down the long hallway toward the door of Small Class No. 1.

Little Pumpkin stood alone, still waiting to be called.

4

NO EXCEPTIONS TO THE RULE

We are a Chinese school. You chose to place your child here, and you must conform to our educational style.

—TEACHER CHEN

Rainey is the kid who plays tricks on his family, launches impromptu a cappella sessions, and organizes kids in play. I love that Rainey's personality sticks in the memory of most people who meet him, and I've always welcomed displays of individuality and playfulness. I want my child to understand what it means to color outside the lines.

Yet Chinese culture promotes conformity: The nail that sticks up would be hammered down, the bird whose head sticks out easily shot, and the tallest tree easily destroyed by the wind, as the proverbs go.

What I was learning of Chinese teachers' methods stunned me, and I was shaken by the thought that Rainey might have his tiny spark extinguished at school. At home, our little boy liked to laugh uproariously when Rob or I did something funny, doubling over or slapping the floor with his hands, continuing to chuckle past the point when a joke lost its humor. For our fun-loving boy, the physicality of laughing often became the joke itself.

Were Rainey's teachers prone to screaming, shaming, and threat making? Were they embarking on the long march to make Rainey conform to his environment? The qualities I was observing in the teachers I'd met were clearly part of an authoritarian teacher culture, but were there kinder and gentler gradations on that scale?

"Will the police take me away if I don't nap?" Rainey asked Rob one Saturday, as he was going down for his afternoon snooze.

"Why would the police come if you don't nap?" Rob said, tucking a blanket around his neck. But Rainey didn't answer.

The following week, on a lazy weekend afternoon, Rob and I watched as our little boy curled up in a fetal position on the living-room floor. This "baby Rainey" purposefully squeezed his eyes shut, squinting with the effort, as if trying to block out some kind of apparition. After a few seconds, he cracked open his eyes to peer surreptitiously at his environs. He did this several times, until I finally got it: He was trying to escape detection.

Detection from *whom*?

"Baby Rainey" suddenly jumped up to standing and swaggered about the room. Immediately, I realized my son was imitating a teacher.

"You must sleep! Close your eyes and rest. If you don't, I'll call the police," Teacher Rainey boomed, wagging a finger over the spot where "baby Rainey" had lain.

It was clear what this meant.

"Yes! Close your eyes!" Teacher Rainey boomed. "If you don't, I'll call the police to take you away!" When I was a child, American culture presented the police officer as a friendly authority figure, a helper of elderly women crossing streets, and a source of fascination for young children. In Chinese culture, the police are often used as the means to an end: coercing children to do what an adult wants.

"I said close your eyes," Teacher Rainey boomed again. "If you don't close your eyes, I'll send you to *tuo ban*. You'll never see your classmates again."

Tuo ban was the class level for the two-year-olds. The teachers were threatening to demote our son? Rob and I looked at each other, stunned.

As my father's daughter, I'd heard many threats growing up, and my aunts and uncles also thought nothing of tossing off warnings to instigate behavior: "If you don't study hard, you'll grow up to be homeless." "If you eat too much chocolate, you'll get fat." "If you don't become a lawyer, you'll be poor." (For my sister, they substituted "doctor" for "lawyer.")

The Chinese I know can be very specific about their threats, especially when it comes to naming consequences: They are world-class experts at fear-based motivation.

At home, I'd explained to Rainey that brushing teeth kept the dentist away, and the following day I overheard our ayi: "If you don't brush your teeth," she told Rainey, ominously, "insects will sprout from the filth and devour your face while you sleep."

For the next several nights, Rainey slept in fear of face-eating bugs, and I instructed Ayi to leave oral hygiene to me. The police threat, however, didn't seem to concern our little boy. Was it because he'd heard this one too many times in the classroom?

Rainey was either unwilling or unable to confirm the source, so I did what any self-respecting American parent would do.

I requested a parent-teacher conference.

* * *

TEACHERS CHEN AND CAI stared at us from their pint-size chairs.

Their students had been tucked away in another room for nap time, and the homeroom of Small Class No. 4 felt suddenly cavernous and cold, as if its beating heart had been suddenly snatched by a surgeon's forceps.

Rob and I were seated in child-size chairs of our own, rears low to the ground and knees propped awkwardly. I got straight to the point.

"Rainey doesn't like to come to school," I said.

"Ah," both teachers said, nodding, as if this were no surprise. Teacher Chen spoke first.

"Before Rainey came here, which school was he attending?" she asked.

"Happy . . ." I stopped and cringed with embarrassment. "Um—Happy Kids." The previous year Rainey had attended a nursery in Shanghai owned by a Canadian and run by a Frenchwoman. The Chinese would balk at naming a school for joyful contentment; Wisdom First, Sacrifice Is Golden, and World's Best Math is more like it.

"Ah, Happy Kids," Chen repeated, nodding, as if a hunch had been suddenly confirmed. "Were the classmates foreign or local?"

"Mostly foreigners."

Chen nodded again, diagnosis complete. "We think that Rainey is very smart, enthusiastic, and warm, and willing to learn," said Teacher Chen. "I think the biggest problem he has is with the *guizhi*—the rules."

"What rules?" I asked.

"He feels restrained. We think he has a problem with discipline."

"Could you give me specific examples?" I asked, a tinge of anger creeping into my voice.

"Easy," Rob whispered, under his breath, from his tiny chair, knees nearly up at his shoulders.

Chen and Cai exchanged words in Shanghainese and finally turned to us again.

"At . . . Happy Kids," Chen said, "he experienced a foreign education culture. Western education culture is much more . . . *casual* than Chinese education culture."

I sat. I stared. Chen offered an example. "If Rainey jumps around and falls, the American mom will think, 'No problem. Kids will be kids.' But for the Chinese mom, it's very important that they don't fall down. Rainey has not learned that school is not fully for enjoyment."

"What has he done? Give me an example," I insisted.

"He goes down the slide headfirst," she said.

"We think going down the slide headfirst is good," I said. "He's experimenting, having fun."

"Ah!" Chen said knowingly, as if suddenly realizing the mother, not the child, was the problem. "But we have twenty-eight children in the class. If they all do it that way, it will be dangerous," Chen told me. I recalled that the teachers' directives in the Child Development Book indicated a world full of danger: "Change clothes often to minimize contamination. Drink cooled, boiled water if possible. Don't go to public places too often. Wipe floors with a wet towel. Exercise, but not too strenuously."

"The first thing is," Chen said, "you have to observe safety rules. You can't run too fast, you can't bump into people. You must finish your lunchtime meal. You must listen to the teacher." Cai sat next to her, the silent but nodding lieutenant. My relationship with Cai had never recovered from the botched Coach purse delivery, and when she wasn't chilly toward me, she was dismissive.

"We understand," I said, guiding the conversation back to the point at hand. "But Rainey says he was told that the police would take him away if he didn't nap at nap time. We are concerned about your use of threats as a tactic."

Chen and Cai looked at each other for a beat and whispered to each other in Shanghainese. "It's not allowable to threaten children, no matter at home or abroad," Chen said, finally. "We have never told children we would call the police."

"They're flat-out denying it?" Rob muttered to me. "Where else would Rainey have heard it?" He took over.

"Rainey says he's scared to come to school because he's told the police will come if he doesn't sleep," Rob announced.

"I've never used force," Chen said, neatly skipping over the police threats.

"Why do *you* think Rainey isn't liking school?" I asked. And gradually, I unfolded all my concerns about the dictatorial classroom environment: forced naps, near-constant lining up, a lack of free expression during art.

Chen had a functional response—a reason—for everything. Yes, the children must lie ramrod straight during nap time, because the beds are close together and others will be disturbed if anyone fidgets.

Yes, the children line up for bathroom breaks, otherwise chaos will ensue. Yes, water is only drunk at designated times, so as not to interfere with instruction time. Yes, the children must learn the fundamentals of drawing before they can experiment. No, the children are not allowed to talk during lunch.

"Not at all?" I asked, visions of my very lively elementary school lunch table in Houston dancing in my head. "No talking while eating?"

"Not at all. We're not like other Western schools where everything is very . . . *casual*," Cai said, switching to English for the last word. "We need them to finish eating on time. Also, if they're talking they might choke on their food."

"What about sitting in chairs for a long time?"

"Yes, they must sit straight in chairs. For music class it makes their voices clearer," interjected Cai.

"What about discipline? What if a child does something wrong?"

"I'll talk to them and ask them to think about it for a while."

"Will you take them to another room and ask them to stay there alone? Or make them stand outside the classroom?"

"No, never," Cai said. "I don't think it's acceptable to threaten children, whether at home or abroad in the West."

"But someone did," I told her, recounting what I'd seen Rainey recreate at home.

Perhaps it was the classroom ayi, Teacher Chen said, her tone taking on an edge. "I will talk to her. You know—we *are* trying to adopt some Western ways." Chen and Cai informed us that the government is working very hard to reform traditional education.

Chen turned to me with a question. "We have a problem of our own to bring up," she said. "Rainey has a habit of straddling other children and pretending that they're domesticated riding animals." She brought her hands up to her chest, fingers curled over, and bounced twice in her tiny chair—*beng, beng*—as if she were riding a bucking donkey. "There is an imaginary bridle and whip," Chen explained, "and he'll try to make the other children gallop."

"Oh," I responded, as flatly as I could. Chen looked at me with surprise, as if she'd expected me to recoil with horror. I was concerned, but frankly, I also wanted to laugh at Chen's pantomiming.

"The other children complain to me," Chen added for good measure, with another two bounces. *Beng, beng.* "When we first noticed the behavior, we'd try to talk to him about it. But now all we can do is yell."

"You can't argue with that," Rob said to me in English, under his breath. We sat perched in our chairs, two parents of a naughty child, humbled and mute.

Chen spotted her opportunity. "We are a Chinese school. You chose to place your child here, and you must conform to our educational style." She seemed to be speaking directly to me, and I suddenly knew Principal Zhang had told her of my attempts to weasel into a classroom for observation.

I whispered to Rob in English, "Does she think we're running a jungle at home?"

"Well, if Rainey is riding his classmates—then, yes," Rob muttered back.

"If you let him run around very casually at home," Chen admonished, "he comes to school and tells the teacher, 'Well, Mommy says it's okay to do this.'"

I nodded.

Chen looked at me, specifically, again. "He needs to think that Mommy and teachers are on the same side. You decided to send him here. You need to trust our educational style, and you must do the same at home."

I can't pinpoint exactly how we knew it was time to leave, but Rob and I rose from our chairs at precisely the same time, and Chen and Cai mirrored us. We'd stepped into their classroom expecting an honest conversation, but the rules of *mianzi*, or face, didn't allow them to own up to anything. I'd been naive to expect that a direct challenge would be effective.

I bowed my head. "It's time for you to have lunch," I proclaimed.

Because protocol required them to throw some face our way, to acknowledge my reason for calling the meeting, Chen promised to "look into" the police threats.

I knew we'd never speak of it again.

* * *

AT THIS POINT in Rainey's journey, I suspect many of my American friends might have sprinted through the black gates of Soong Qing Ling, kid in tow, without a backward glance. Some might have told off Teachers Chen and Cai while perched in their tiny chairs. Others might have fled to the nearest international school, whose philosophies ranged from Reggio Emilia to Montessori to Waldorf. One friend had recently pulled her child out of Soong Qing Ling after trying on the Chinese option as casually as she might try on a *qipao* dress on for size at the local fabric market.

I was certainly incapacitated some days by doubt and fear, but I also wasn't ready to abandon the mission.

Many foreigners living in modern Shanghai were brought in by multinationals, law firms, and service agencies eager to tap the immense Chinese market for profit. When these Americans, Brits, French, Germans, and Japanese weren't toiling inside office towers, they went on trips to Thailand or Bali, took escapades in their chauffeured nine-seater vans, or dined out at European and Mediterranean restaurants. In other words, they tried to insulate themselves from the experience of China as much as they could. A couple of American women I knew rarely left their apartments and villas unless their spouses' company-appointed drivers waited at the end of their driveways, doors ajar to air-conditioned protective pods as they traversed this megacity.

Rob and I wanted to be another type of foreigner in China altogether.

Rob's time in rural China as a twentysomething had acquainted him with the country and its culture. During his time teaching with the Peace Corps, he'd befriended Chinese students who were polite and inquisitive and respected education, and he carried with him a certain comfort with placing his own son in the Chinese system.

Meanwhile, my own heritage made the Chinese and their behaviors immediately familiar; it was as if I looked into Teacher Chen's eyes—no matter how harsh and authoritarian their glint—and immediately recognized my father's intentions (sometimes misguided but always well-meaning). Despite all my struggles, I never lost sight of the value in discipline and hard work delivered the Chinese way.

Rob and I ultimately traveled circuitous paths to arrive at the same point: Philosophically, we saw value in the Chinese approach to discipline and academics, and we also wanted our son to experience the same culture we'd grown to appreciate. Today, Rob and I bike in the sweaty tangle of Shanghai street traffic, eat fiery Hunan food, and journey to the far corners of China and Southeast Asia during holidays. We're Americans, but our holistic identities are not linked to any particular place. "You're a global citizen," my aunt Kari told me. The term was overused, but in many ways she was right.

We'd chosen to spend our son's early years in China, in part because we felt it was important for him to learn another way of life. We wanted a nimble child, a boy who could handle a world that was uncertain and rapidly changing, with the confidence that he could find his place within it. My aunt Liang, a career psychologist with a focus on ethnic culture, challenged my fear and doubt most starkly: "Why are you stressing over this 'local school' decision? You live in China. Why would you expect Rainey to attend anything other than a Chinese school?"

Rainey's Teacher Chen was right. We'd *chosen* to put Rainey into a local school. We'd elected to immerse our son in the culture, hoping he'd absorb the language and some of that renowned Chinese discipline. The problem came when we expected to cobble together an educational experience as if we were plucking items off a menu. I wanted Rainey to learn Mandarin, but I was uncomfortable with forced nap time or egg eating. We knew he stood out as a foreigner with his brown hair and odd mannerisms, but we'd expected his teachers and classmates to override basic sensory cues and accept him fully as a peer. (I'd once caught a teacher calling him *xiao laowai*, or little foreigner, which, to my surprise, didn't seem to bother my son.) I'd exuded arrogance toward Teacher Chen, who had years of experience in China and a master's degree in education, while I had none.

Why did we expect to gallop into a Chinese school and bend its educational culture toward our pole within months of arriving?

What gave me the right to feel entitled to *an exception*?

I found comfort in the fact that Rainey's school was a government-designated "model school" with access to special funding and privileges. (By contrast, Little Pumpkin's school was average, in a nondescript suburb of Shanghai.) Soong Qing Ling strove for smaller class sizes and teachers with master's degrees. Seventy percent of its teachers had certificates specifically granted for early childhood education, and they went on frequent trips abroad to study other school systems. The physical education teachers spent time in Australia, that bastion of sun and sports. The school's parent population was well-to-do, well-traveled internationally, and exposed to the ways of Europe and the United States.

"We believe in a kinder, gentler education," Principal Zhang had told us during an orientation speech. "We are one family in this school—one united family. Teachers and parents should learn from each other." Soong Qing Ling was at the forefront of a national effort to change China's approach to education, the principal told us. The school is a "testing ground for the future of Chinese education, and in recent years our school has influenced all kindergartens in Shanghai as well as across the nation."

This gave me hope. Anyone with even a casual knowledge of Chinese education knows that academic pressure got very intense during the teenage years, and the news media often published stories of student suicides leading up to important exams. Yet here was evidence the government wasn't entirely satisfied with its education system. Just as I had questions, so did they.

I also wasn't convinced Rainey's teachers were as horrible as my worst fears suspected. On good days, I was captivated by the idea that these government change efforts might offer some upside to the downside of Chinese education and that, indeed, Rainey might come out of his schooling flexible and gritty, as well as some kind of academic superstar to boot. I planned to step up my efforts to find out exactly where Chinese education was headed, and I knew my reporting would better inform our decisions about Rainey's schooling.

As a safety measure, I placed Rainey's name on the waiting list at an international kindergarten down the street from our house. These foreign-run bilingual schools were the closest approximation we might get of a public school in America, but they were prohibitively expensive: up to $40,000 a year. International school was an extravagance Rob and I couldn't afford without making "adjustments," but exploring a backup plan seemed something a responsible parent should do. Anyway, it would probably take months for a spot to open up.

Meanwhile, Teacher Chen had made clear our mission for the weekday hours of eight a.m. to four p.m.: We'd need to ensure Rainey didn't mount his classmates like they were horses, inure him to a culture of compliance by threats, and help him adapt to his foreign environment.

* * *

AS PART OF that promise, I hired a Mandarin tutor for Rainey.

It's commonly said that Chinese is one of the most difficult languages in the world to learn. The Chinese language consists of more than forty thousand distinct characters, with each appearing to be a random assortment of straight and curved strokes squiggled in ink, with no immediately discernible order.

This inscrutability is complicated by the fact that spoken Mandarin has four tones, so that any phonetic sound, such as "ma," will have a different meaning depending on whether it's spoken with a steady high tone, a rising tone, a falling *then* rising tone, or a falling tone. Thus, even if you manage to commit thousands of characters to memory, you still must hit the tones in everyday speech.

"The easy characters to remember are the ones that resemble what they mean," said Rainey's tutor, a young preschool teacher I'd found in the neighborhood, who was excited for a little extra cash weekdays after work. She began to break the task down into discrete parts. The character 火 means fire, she told Rainey, and you might say it resembles flames bursting from two sticks of wood. When you extrapolate from there, she said, you'll see that characters containing the radical 火 might typically have a meaning related to fire. For example, 烧 means "burning," 烤 means "toast," 烫 means "something extremely hot." Such radicals, or root forms, offer clues to sound and meaning.

The tutor would draw little pictures to help. For example, 山 meant mountain, and 上 indicated "up." Other little tips and tricks assist the learning process, but ultimately, no Chinese teacher will deny that memorization and drilling of thousands of whole characters is ultimately a large part—if not entirely the essence—of learning Chinese. No child can escape that process which has become a "dirty word" in American classrooms: rote memorization.

Learning Chinese takes years. Seven or eight years to learn to read and write three thousand characters, estimated the linguist and Sinologist John DeFrancis, while students of French and Spanish might achieve a comparable level in half the time. Anyone trying to learn Chinese "will always be frustrated by the abysmal ratio of effort to effect," wrote the linguist David Moser. Moser writes with humor of an incident where he couldn't conjure up the word "sneeze." He happened to be dining with three PhD students at Peking University. "Not one of them could correctly produce the character," Moser wrote. "Now, Peking University is usually considered the 'Harvard of China.' Can you imagine three PhD students in English at Harvard forgetting how to write the English word 'sneeze'?"

My own journey learning Chinese had been tortuous. My parents sent me to Sunday Chinese school in Houston, and I absorbed as much as I could, that is, when I wasn't hoping for a sudden bolt of lightning to strike me dead as I traced out characters in my notebook.

In China, schoolchildren are anything but casual Sunday learners. They drill daily, and they're required to reach full literacy astonishingly quickly. "A child's ability to memorize is very good at this stage, and it should be tapped," a primary school teacher told me during an interview.

China sets rigorous curriculum standards, with no leniency for even the youngest children: First- and second-graders should recognize 1,600 characters and write 800 of them from memory. By fourth grade, the level is 2,500 characters, and by the sixth grade it's 3,000 characters and writing almost as many, although many schools' individualized curriculum actually calls for more. Full literacy requires an astonishing 3,500 frequently used characters to be committed to memory, according to the Chinese curriculum standards for full-time compulsory education. That means a typical Chinese first-grader—six years old when they start the school year—attends hours a week of Chinese class, reading, writing, and reciting every day. (The process itself drills rigidity and memorization into a child's routine; some education watchers say it's the unforgiving task of learning Chinese itself that lays the groundwork to killing curiosity and creativity in the Chinese schoolchild.)

Ironically, as fearsome as the task seemed to the intellectual mind, Rainey was having more fun memorizing Chinese characters than learning to read in English at this early stage. He'd easily learned his ABCs, but having to remember the phonetic sounds that accompanied each letter, not to mention consonant combinations such as "sh" and "ch" and "ll," made him throw up his hands in despair.

Some days, his Chinese learning journey was gratifying.

"Carrots are for rabbits—and I am not a rabbit," he said to me, in Chinese, laughing. I'd coaxed him to eat this vegetable at dinner, and

I'm sure Rainey loved being able to reject my nutritional efforts in Chinese as well as English.

Other days were challenging, such as the time Rainey refused to invite Chinese friends to his birthday party.

"I don't want to speak Chinese on my birthday," he said.

"Don't you speak Chinese in school?" I responded.

"Yes, but I'm not at school right now," he replied. Good point.

Another time, he walked away from Ayi midsentence, leaving her staring after him in puzzlement. He explained to me and Rob, in English. "I'm tired of speaking Chinese," he said. "Also, paleontologists don't need to speak Chinese." Rainey had decided upon turning four that he wanted to get paid to dig for fossils when he grew up.

"Paleontologists *do* need to speak Chinese," Rob replied. "They find a lot of dinosaur fossils in China."

Good point.

⋆ ⋆ ⋆

WHEN THE FACTORIES in the Yangtze Delta work from morning through night, and weather patterns are stagnant, a perfect confluence of factors makes the air a dense, dark, choking fog.

Shanghai is home to a thriving steel-manufacturing sector and a multitude of petrochemical, plastics, rubber, metals, and machining plants. These factories and their bellowing smokestacks infiltrate and encircle Shanghai, and some days, thick blankets of industrial smog obscured the morning sun and shrouded the skyscrapers outside our window in a cloak of gray.

This was China's pollution problem, which presented a unique challenge for us.

Rainey suffers from coughing fits, which worsen during times of high pollution, and a doctor cautioned us he might be showing early signs of asthma. Many children grow out of it, the doctor told us, but in the meantime, we should keep a rescue inhaler with him at all times. Including, he told us, when our boy is at school.

As part of the doctor's prescription, we'd have to stay inside on particularly smoggy days. Checking the air quality each day was easy: Each morning, sixteen floors up, I would part our dining-room curtains and gaze north over the red rooftops of the *shikumen* buildings that sat low to the ground, past high-rise apartment towers, and over Yan'an Highway, already choking up with Baojuns and Buicks on the daily commute.

I was seeking out my personal pollution monitor: Jing'an Temple in all its golden glory. Otherwise known as the Temple of Peace and Tranquility, the temple sits about half a mile from my sixteenth-floor window. Dating back to AD 247, Jing'an Temple was relocated during the Song dynasty, renovated during the Qing, destroyed in 1851 and then rebuilt, and today it's an imposing structure of gold leaf and layered rooftops that ascends into the sky.

On clearer days, when I could make out the fine detail on the gold-layered knob at the temple's top, the air-quality index usually notched in below 100. "Safe for most people," declared my mobile air-quality app.

If the temple's edges disappeared into a gray-black haze of smog, like a photograph whose borders had been deliberately smudged, the index notched in between 100 and 200, already well outside the level deemed safe by the US Environmental Protection Agency: "Unhealthy. Restrict outdoor exertion."

If I saw *nothing* of the temple, much less the buildings the next block over, I knew the index was at least 200, 300, or above: "very unhealthy" or "hazardous." Time to hunker indoors in our hermetically sealed apartment, mini-refrigerator-size air purifiers cranked to high.

One week in December, I parted the curtains and looked north past Yan'an Highway. No Jing'an Temple. As my eyes began to focus, I realized I couldn't even see the buildings the next block over but for a faint shadow here and there, phantom outlines of skyscrapers shrouded in smog. I couldn't make out the sun. The index had surpassed 500: so toxic, it exceeded measurement capabilities. (Los Angeles and London usually hover around 50.)

"Airpocalypse!" screamed a Western newspaper, while Shanghai officials ordered all schools to suspend outdoor activities.

"I can't breathe, so I'm staying in," a friend texted me, canceling a planned coffee date. "We're keeping Rainey home," Rob told me, teeth gritted. "Windows closed, filters on high." For days we cowered indoors, taking our stand against World War Smog. The pollution finally lifted after a long, interminable week.

I returned Rainey to school with a newfound determination. Asthma isn't a medical condition most Chinese are familiar with, and I'd been hesitant to call attention to any feature that would brand my son an exception to the rule. But it was time to talk to Teacher Chen.

"What is that?" Teacher Chen said, glancing at the blue plastic-and-metal contraption in my hand. I'd pulled her aside at pickup time, armed with the Chinese vocabulary for asthma, emergency, and rescue inhaler.

"An inhaler—*xiru qiwuji*—medication to help Rainey breathe," I told her. "He has asthma. If you see Rainey having trouble breathing, could you please give him two puffs?"

"We don't keep any medication here. Take it to the school nurse," Chen said.

"*Buhaoyisi*—I find it embarrassing to say," I said. The Chinese use this nicety to preface anything troublesome, contradictory, or confrontational, and what I was about to say classified as all three. "*Buhaoyisi*, but a fast response is important, and the nurse's station is too far away. Could we keep the medication near Rainey?"

"We don't keep medication in the classroom," she repeated, and she turned away.

"How about the coatroom next door?" I asked. But Chen shook her head without turning around. A backside in retreat—that summed up my rapport with Teacher Chen.

The nurses' station sits in the same yellow-and-white stucco building as the guard's glass house at the school entrance. Stretching along the side of the building is a long trough of opaque, black sinks, exactly waist-height for a child. Above the sinks sit an array of golden plaques, which together declare Soong Qing Ling a model kindergarten and a "laboratory for infant education."

This area had a distinct purpose in the mornings: the daily *tijian*, or health assessment. Children would stream through the gates past the guards, lather and rinse their hands at the low, black sinks, dry off, and then step into one of two long lines that snaked into the nurses' station. At the head of each line, a pink-uniformed nurse peeked into tiny mouths and turned over palms to check for the red dots that signified hand, foot, and mouth disease.

Depending on her assessment, the nurse would hand each child a thin, colored tab of plastic the size of a USB thumb drive.

"Let's go, Mom!" Rainey would say, clutching the colored tab as we walked together up to the classrooms. A red tab indicated good health: no action required. Blue signified to the teacher that the student was scheduled to take daily medicine. Pink—the color of Pepto Bismol—indicated stomachache. Yellow meant the child showed signs of cough. Green was a mystery, so I asked Rainey.

"That means bleeding or hurt," he explained one day, smiling up at me as we traversed the huge grassy lawn in the center of campus, which I came to call Big Green. The daily check-in process was orderly, completely silent, and an effective way to communicate between nurse and teacher: Color-code the missive, and dispatch children as the messenger. Chinese efficiency in processing large crowds always amazed me.

One day Rainey got a yellow card, which meant "cough." And the day after that, and in fact every day that week. But he was perfectly healthy—he should have been getting red! Something was going on.

"What does 'yellow' mean?" I'd asked him, as casually as I could. He clutched the yellow tab as if it were a prized toy.

"It means 'cough.'"

"But you don't have a cough this week," I'd said. Rainey didn't answer.

The next morning, I positioned myself so that I could observe the head of the nurses' line, and to my surprise, I saw Rainey carefully cough—just once—right as the nurse was peering into his mouth.

Yellow card.

Rainey and I made our way up to the classroom. "What is different when you get yellow?" I'd asked him, feigning nonchalance.

"More water," he'd said.

The kids don't get as much water as they want? I'd thought with a smidgen of horror. This scene replayed in my head as I approached the nurses' station that day, armed with Rainey's asthma inhaler. I stepped inside.

"May I bother you a moment?" I said, as Nurse Yang Cun Cun approached, sporting a light pink frock and dark brown contacts whose irises streaked blue as they approached the whites of her eyeballs. A woman with broad shoulders also joined us from a back office, and we three installed ourselves in tiny chairs around a low table.

"Rainey suffers from *xiaochuan*, and I'd like to talk to you about keeping medicine in the classroom in case he has trouble breathing," I said.

"The teachers can't keep medicine in the classroom," Broad Shoulders said, as if repeating Line 37 from some kind of Operating Manual. "We'll keep it here." She motioned to a yellow cabinet against a side wall. Through the frosted-glass windows, I made out the shapes of tiny bottles and white cardboard boxes the size of my palm. Nurse Yang sat beside Broad Shoulders, acknowledging her lower rank with her silence.

"*Buhaoyisi*—I find it embarrassing to say, but that will be a problem," I said. "We need to keep it close to him in case he has an attack."

"The teachers can't keep medicine in the classroom," she said, repeating Line 37.

"Is there another possibility? How about in his student's shelf in the coatroom?"

"We can't keep medicine outside the nurses' station," she repeated. It seemed a reasonable rule, but in this case, the distance from Rainey's classroom was a problem, and we needed to solve it.

"The medicine is not dangerous . . . it won't affect the kids who don't need it," I said. "It only opens up the airways for those who cannot breathe."

"We'll keep it here," Shoulders said firmly, staring at me. China was a place of many rules, and in this case the official rule was clear.

But where was the work-around? I thought. The Chinese swam against so many spoken and unspoken rules—governmental, societal, cultural—that simply to function, often you had to navigate the system, even if it meant breaking the law. The rules, taken individually or in aggregate, were simply too restrictive to follow to the letter of the law. An ancient Chinese proverb **制度是死的，人是活的** proclaims "Rules are dead, the people are alive." An acquaintance who works at a not-for-profit told me that "the rules here are either too strict or too stupid, which ultimately makes them just for decoration."

Whether decorative or real, here I was bumping up against one of these rules. The Chinese always seemed to know just how to resolve such conflicts through *guanxi*, gifting, an appropriately timed offer, an uttered phrase, or some other dance move. But after a few years in China I still found it all as mysterious as a peacock mating ritual. I may be of Chinese ancestry, but during most of my early days in Shanghai, I felt very much the bumbling foreigner.

Unlike me, I could see that Rainey had adjusted. He was finding his own way to get things done. He wanted more water, and he'd discovered that faking a cough was the most effective way to accomplish his goal without triggering the teachers' ire.

His mother had no such magical work-around. I requested parent-teacher conferences. I confronted Teacher Chen about medical issues. And here I was, challenging two school nurses, inhaler in hand. I was utterly without a compass; I could only repeat myself.

"If we keep the medicine here, it's not close enough to help him," I said.

"We can get to him in three to five minutes. We'll hurry," Shoulders said.

"It will be too late," I said, my tone getting insistent. "If it's an emergency, he won't be able to breathe. He'll immediately need two puffs."

Shoulders spotted her opening. "On days that he is coughing he'll rest in the nurses' station. We'll sit him out during outside play," she said, motioning to a small cot nearby. I glanced at the thing, a pitiful little bed tucked into the corner of the room.

"No, no, no, NO!" I said quickly. "Running is good for him because it helps his lungs expand and contract."

"Well, then we'll just keep the medication here," Shoulders said. Line 37 of the manual, again.

"You won't be able to get there in time. This is what will happen," I cried, desperate. I put my hands up to my throat, mustered the most contorted face I could, and started gagging and choking. "Aargggh, eeeeee-gghh, elllhhhh," I grunted, as my tongue wagged and my eyes bugged.

"Oh!" Nurse Yang exclaimed. Broad Shoulders and Yang shifted in their tiny seats.

"*Xiaochuan* can kill people if they don't get relief. You would let him die?" I asked, going in for the kill.

"No, no," they said, looking at each other. I kept my hands clasped around my throat. I sensed a breakthrough was near.

Broad Shoulders arrived at the solution first. "Perhaps Rainey needs to be transferred to another school, a school for children with special needs," she said, with an authoritative bob of her head. "We can give you a referral."

"Oh! No, no, no," I said, quickly, removing my hands from my throat. "No, no. Rainey's condition is not serious." I'd heard of such places. They were institutions where "problem" children—those with anything from autistic spectrum disorder to multiple sclerosis—were neatly tucked away from society. In general, China's school system lacked a formalized method of identifying children with disabilities or special needs, and there wasn't a practice of trying to keep these children in the mainstream (although I've met a handful of educators and activists working to change this).

Broad Shoulders and Nurse Yang simply sat there. They looked at me. They waited.

"Let's just keep the medicine here," I said, handing over the inhaler. Broad Shoulders nodded, and the inhaler disappeared into her hand. I would talk to Rainey's doctor about managing his coughing with a daily steroid inhaler, a longer-term preventative therapy we could administer at home.

Broad Shoulders wasn't finished with me yet. "We need a paper from the doctor with the child's name on it and the name of the medication," she said.

"I don't have one," I said, without thinking. "I got the medication in America. This paper you speak of—I handed this over to the pharmacy."

Shoulders only repeated her request, unblinking, as she gazed at me. It was clear that Rainey had begun to adapt to his top-down environment, yet his mother was still struggling under the system's authoritarian ways. Instinctively, I knew these conflicts might eventually force some kind of decision about whether Rainey should stay in the system. But in the meantime, it was clear that if I didn't conform to Soong Qing Ling's directives, my son would be kicked out of school.

"I will bring the paper tomorrow," I told Broad Shoulders, looking down at my shoes.

5

NO REWARDS FOR SECOND PLACE

Competition motivates me. Rankings motivate me.
It pushes me to action.
—DARCY

Rainey was born in Los Angeles, where year-round sunshine graces park outings, lactation consultants make house calls, and daycare providers believe in infant choice, free-form play, and locally sourced juices of cucumber and carrot. Our California playdates were typically sun-kissed affairs, with parents sipping pinot noir and paring down a wheel of Brie as children wandered around a grassy park.

I found the Chinese equivalent less relaxing.

There are few good neighborhood parks in the bustling megacity of Shanghai. Land is simply too valuable, and local government too pressed to spur growth, to develop an open space purely for enjoyment. So, near the end of Rainey's first year, we were invited to meet half a dozen of his classmates in the basement playroom of a seven-tower apartment complex, each tower stretching twenty, thirty, or more stories into the sky.

The entire affair started with an inquisition and ended with a foot-race.

"What's Rainey's birth date and height?" Mom One asked me, five seconds after I arrived. I answered, and she immediately launched a rapid mental calculation.

"That makes Rainey the second-oldest in Class Number Four," she told me.

"My son is *the* oldest," Mom Two jumped in. "His birthday is September twenty-seventh, so he's older by . . ."

". . . two weeks," said Mom Three.

"Yup, and Rainey is the third-tallest in the class," Mom Two declared.

I nodded, slightly stunned. Did she have all these numbers in her head?

Next, the mothers maneuvered the children into race position. In all, there were five boys. One mother lined them up at the edge of the room, an expansive space carpeted in bright orange and gray. I glanced over at Rainey and was surprised to see that he slotted himself into place. He was on all fours, nose pointed toward the opposite wall.

One mother lifted her arm dramatically, then dropped it. "*Yibeiqi!*" Ready, go!" The greyhounds were off.

The children crawled, arms and legs cycling rapidly as their little bodies hurtled toward the opposite side of the room. I'd never seen my son move so fast, his feet up in the air as he lurched forward on his hands and knees.

"*Jiayou, jiayou!*" Add oil! Add oil! chimed the voices of several mothers, using a common Chinese cheer. The woman who launched the race slapped her hands together, making piercing sounds.

"Go faster! Go faster!" One of the mothers clapped beside her boy, running the entire distance in her pumps as he crawled.

One little boy fell behind immediately, and by the midpoint he simply stopped, splaying his arms and legs out flat as he plopped his belly onto the carpet. He lay there, carcass spent, for the remainder of the race.

"*Jiayou, jiayou!*" Add oil! Add oil! The voices were cacophonous, and they echoed throughout the clubhouse as the remaining boys continued on.

When the winner reached the wall, he immediately turned around, watching the rest of the pack advance on his tail. Rainey came in second. The boy who touched the wall third began to cry. Clapping ensued as the mothers cheered the end of the race.

"One hundred points for *Tian Tian*! One hundred points for *Tian Tian*!" one mother said.

"They're so competitive with each other!" one exclaimed, laughing, as several women rushed up to congratulate the winner.

Rainey sat with his back against the wall. He was neither joyous nor upset. He was just taking it all in, his first Chinese playdate.

* * *

AS I PONDERED what about that gathering had so troubled me, I realized it was the American in me that declared self-esteem a holy grail in parenting and education.

Americans generally hold a child's regard for self—an emotional evaluation of one's own worth—with feather-soft gloves as if it were a panda cub (which requires special care because it is blind at birth, immobile and unable to feed itself). A child's self-esteem is treated almost as a physical organ, nearly as important as the beating heart, which pumps blood to the brain, organs, and extremities. Self-esteem is the concept that compels adults to give every child a trophy, no matter if they finish dead last in the neighborhood bike race.

Along this vein, American parents generally refrain from openly comparing children, in line with a culture that values the individual (and her feelings). It's impolite to utter which child is smarter, better, faster, or shows leadership qualities that others don't have. (If you must compare children, find a circle of best friends and whisper over Thursday-morning coffee.)

The Chinese have a more utilitarian approach.

"Self-esteem" doesn't exist in the Chinese lexicon, at least not in the way Americans use it. In China, a child's regard for herself is rarely as important as a stark evaluation of performance. Almost as if childhood were an Olympic sport, the Chinese rank children on everything from work ethic to Chinese character recognition and musical skill.

Comparisons can be informal and conversational.

"He's not as smart as his brother, but he's a better singer," my acquaintance Ming said to me once, nodding at one of her boys, in earshot of the less-smart brother. Sometimes the desire to rank is combined with a threat. "Does your father love your brother more?" a Chinese teacher once asked my friend Rebecca's daughter. The question came after the girl had a bad showing on an in-class assignment. One time Teacher Chen told Rainey, in front of all his classmates, "Your Mandarin is not good."

Rankings can also be formal, and public, and in a Chinese education the comparisons start early. At Soong Qing Ling, the monstrous bulletin board that sits outside of Rainey's classroom was the home for public ranking for Small Class No. 4.

Big Board might post teachers' assessments of each child, a report card displayed for all to see: Who clocked in timely arrivals at school? Which child greeted the teacher with a smile? Who finished every lunchtime grain of rice? Star stickers and happy faces were pasted next to the names of each child who'd made the grade. Other times, a chart might be offered, with the names and numbers of twenty-eight children in the leftmost column. Each succeeding column listed some kind of data point; the first time I saw the height-and-weight chart, my eye went straight to my son's numbers. In December of his Small Class year, he weighed 16.7 kilograms (36.8 pounds) and stood 105.6 centimeters (almost three feet six inches) tall. It's not surprising that my eyes didn't stop there; it was simply instinctual to begin scanning the rest of the numbers to see how Rainey stacked up. The setup of the chart itself invited comparison and competition.

Only Little Li and Little Wu were taller than Rainey. Just as the playdate moms had said: Rainey was the third tallest in the class. He was four-tenths of a kilogram heavier than Little Hong, and at this, I have to admit, I felt like doing a little fist pump.

The next week, eye exam results were posted, as well as the children's hemoglobin levels; Rainey's was inside normal. A few children clearly suffered from anemia, and a "urinalysis would be forthcoming," the teachers wrote at the bottom of the notice. In the United States, medical privacy advocates might march on the principal's office, but they'd have trouble in China, where the government bans organized protest and the Chinese word *yinsi*, or privacy, had the negative connotation of "having something to hide" until just a few decades ago in the cultural lexicon.

As the months passed into that first year of school, Big Board began to display information that more directly compared performance and ability, as if some Master had declared that it was time to up the ante. With each presentation, parents gathered eagerly, and I could always tell when Big Board posted new information by the number of bodies gathered around, heads bobbing with anticipation. The following year, the Big Board would display prowess at recorder play, for all to see:

The ring finger of Student No. 20 is not stable.

Student No. 30 doesn't cover the old hole while changing to a new one.

Student No. 16 doesn't blow out enough air.

Student No. 3 doesn't cover the holes properly.

Beside Rainey's number, No. 27, the teacher had scrawled the same punishing diagnosis as that for No. 8:

Doesn't follow rhythm.

My son was rhythmically deficient. I knew I'd have to get used to these types of feelings as my child was in Chinese school, so I stood there, letting feelings of shame and regret envelop me. Yet, at the tail end of the wave of emotion came a determination to do better.

"We'll practice harder," I'd declare to Rainey, like a parent making a New Year's resolution.

In a Chinese student's later school years, exam results would be posted, with each person assigned a rank by class and grade level. It never ceased to amaze me that, without fail, the average Chinese student placed these numbers on the tips of their tongues; I'd not yet met a single Chinese schoolchild who couldn't recite his current ranking, and this went for kids I would meet in rural areas, too. "Number sixty-four out of four hundred students in my grade," one high school student told me. Little Pumpkin's Master Teacher Wang recited the numbers for her daughter Cindy: "Sixth in math, fifth in English, ninth in Chinese, and fourth in physics out of 47 classmates. Overall, she's 86 out of 395 kids in her grade." These weren't good marks, and Wang had glanced at the floor.

For my own parents, there was only ever one number worth mentioning.

When I was twelve years old, I'd wanted nothing more than a furry mammal to call my own. Unfortunately, my parents considered the tradition of domesticating animals a little odd, and keeping a pet simply didn't fit their life equation.

My preteen self carefully considered her options. A dog wouldn't pass muster, as they were expensive and required emotional care. Cats made my mother sneeze. Gerbils were small and cheap, and I announced my desire at dinner one evening.

"No," Mom responded, resolutely. "Gerbils are dirty. They smell!"

"A gerbil will take time from your studies," Ba exclaimed.

"We don't keep pets," Mom concluded. When she was my age, she'd just immigrated to America with her family, who spoke little English at the time. From her point of view, if she'd been given a pet rodent at the age of twelve, she wouldn't have worked her way into a PhD, and a professorship.

"Sarah has a gerbil," I offered.

"We're not Sarah's parents," Mom responded.

One summer the mosquitoes and heat yielded a stroke of luck. Via my grandmother, we received news about my cousin Fong: The girl had been invited to play a regional piano concert in Pennsylvania (she'd eventually play a solo at Carnegie Hall as a teenager, the ultimate honor for a pianist). Motivated to action, my parents made an offer.

"Take first place—Number One—and you'll get your gerbil," my mother said. The goal would be a piano competition that drew contestants from all over the Houston area. It was a calculated bribe, designed to coax out as much effort as humanly possible.

At the time, the rivalry between Fong's mother and mine—sisters two years apart who gave birth the same year—was legendary. In Chinese culture, not only is it fine to use competition as an incentive, but even better if the competitor is a relative. After all, a failure against a cousin-competitor dished up proof that your genetics are perfectly fine and that it's a lack of effort that has slotted you for a lifetime of mediocrity.

"Fong plays piano so well. You should practice more like her," my grandmother told me once, egging me on. Fong was also twig-thin whereas I was fleshy, and she stayed porcelain pale during the summertime, while the sun browned my face in a hot second. But I figured I'd focus on one thing at a time.

Like any good Chinese daughter, I'd been playing piano since I was potty-trained. My parents have pictures of my four-year-old self tapping out scales while perched on a piano bench, a phone book propping up my tiny feet—photographic evidence of forced musicality. By my preteen years I'd been practicing at least an hour a day for eight years; to me, piano was an eighty-eight-key beast that sat in our living room and kept me away from Kirk Cameron and *Growing Pains* reruns.

But for a gerbil I could eke out one last triumph.

"You'll play 'Für Elise,'" my mother decreed. Beethoven is much beloved by the Chinese, because, as conductor Cai Jindong explained, the Chinese value those who *chiku*, or "eat bitter," and Beethoven definitely was a man who worked hard. Born a commoner, Beethoven lost a string of paramours to class differences, battled an ongoing string of serious illnesses, and began going deaf in his twenties. Still, he continued to create music and remains one of the world's most influential composers. Beethoven is a classic story of hard luck overcome by effort.

For two months I practiced, sweated, and labored over Beethoven. At night, I dreamed about Beethoven, and during the daylight hours, my mother tapped out time on my shoulder with a ruler while my fingers cavorted like horses trotting on the keyboard. Some days I practiced so much that I was surprised when I looked down to see slender and intact fingers rather than overworked red nubs.

In this way, for two straight months, our little mother-daughter ensemble moved toward the finish line.

I don't remember much about the day of competition. Before a panel of judges and an audience of mostly Asian parents, I got through "Für Elise," but my fingers jerked when they should have frolicked like prancing ponies and plodded where they should have been light as a butterfly's landing. I remember the smugness of the winner—also a Chinese daughter—in the large, raised brown mole that stared at me from above her eyelid. As our eyes caught briefly during the awards ceremony, I thought I detected a flicker of darkness, and perhaps a hint of her own unspoken hell.

On the way home, I pleaded and cried and made excuses. "I'm a good pianist, you know I can play 'Für Elise,'" I whined to my parents from the backseat of our car. "I had a stomachache today."

My parents were silent.

"Can I have the gerbil anyway?" I asked.

My parents did not buy me a gerbil, despite the fact that I practically went on a preteen hunger strike. I plotted to run away from home. I cursed Fong, though she deserved every accolade (as a teenager she developed De Quervain's tenosynovitis in her left wrist, likely from overpractice). Still, no gerbil.

That's how I learned that in the Chinese way, there are no rewards for second place.

Certainly, overemphasizing competition can be dangerous.

"Too many years of competition, and you start to see everyone around you as someone to outrank," Amanda told me. By the end of Rainey's first year in Chinese kindergarten, I'd begun meeting with Amanda more frequently, usually at a downtown Starbucks. I had questions about Chinese education. Teachers Chen and Cai would be of limited use; they thought my son was an animal and that his mother needed reeducation. So I sought out Amanda, knowing instinctively that any Chinese high school student who quoted Nietzsche was a deep thinker.

"Coffee black with lots of sugar," Amanda had told me at our first meeting. "I drink a cup a day now." Just eighteen years old, she'd spent her junior year in high school studying alongside US teenagers, and I laughed at the idea that among America's gifts to her was a lingering caffeine addiction.

The first time we'd arranged to meet, I'd scanned the crowd of young professionals chatting animatedly over lattes at a Starbucks near our home, and my eyes were immediately drawn to a tiny figure whose stone-white windbreaker hung on her waiflike frame. The girl had straight black hair that swung past her waist, and she was bent over an e-reader, reading through lenses with no frames, oblivious to the bustle around her. There she is, I thought, the product of a Chinese education. Amanda was stellar in almost every way valued by the Chinese: a top-ranking student, a Model UN delegate, an exchange student to America. From the outside, she was exactly what I'd expected. The inside, I'd come to discover, would be a surprise.

"Shanghai kindergarten never suited me," Amanda said. "Neither did elementary, middle, or high school, for that matter. I was always the odd girl out."

When I stared at her, I imagined a teenage Rainey or a full-grown Little Pumpkin. "Why?" I asked her from across the table.

"I like reading Proust and Camus," she said. "No one else did." While her classmates memorized Mao Zedong quotes, she preferred reading snippets of Western philosophy. She cared nothing for Chinese pop stars, which her female classmates idolized and chattered about during breaks. There was also the fact that during her twelve-year school career, she'd never been selected *banzhang*, or class monitor, a position of power and influence in the classroom.

"I was always on the outside. It might be hard for you to understand," she told me.

Looking across at her, sipping her Americano with two packets of sugar, I imagined the adult version of a young schoolchild, suffering the invisible bruises of being pushed into a cast-iron mold. Was she the kid who goes down the slide headfirst?

"Actually," I told her, "I think I have an idea."

The picture she painted was grim. After years of being forced to gulp down lunch—a ritual that she says started in kindergarten—she's neutral to food and doesn't view eating as a pleasure. On her first day of preschool she'd cried so violently that the teachers walled her off in an empty classroom. She remembers feeling shame and embarrassment. Then there was the psyche cultivated by years of existing inside a classroom with pressure to conform. "Every time I'm in a crowd . . . it's so weird. I lose my autonomy," she told me. "If there are a lot of people, at a party, I try to blend myself into the conversation, I try to imitate their emotion. I make myself like everyone else in the crowd. I lose myself."

Then there was the feeling of being trapped in a race that never relents. "When I'm not studying, I feel like I'm being lazy. When I *am* studying, I feel that someone is always doing more," she said. It was a mentality cultivated by a lifetime of competition. "There's only one definition of a kid in China. A good student who rises above the rest—I don't know how else to be."

"But something is working," I pressed her. "You say your value system is flawed, that competition has ruined your mind. But if you look at just the academics, you're a star." During her year at that New England high school, Amanda had found her math abilities "three years" ahead of her American classmates, and she soon confirmed her prowess on a larger stage: She'd entered a nationwide math competition with thousands of entrants, and placed in the top fifty.

"Does competition make you a better student?" I asked.

She bobbed her head, slowly. "Maybe. My parents told me, just wait, you'll see. It will all be worth it."

"What will be worth it?"

"The sacrifice. The pressure, the competition," she said. "Feeling like I always needed to work harder." Amanda's family was shooting high. This year she would be applying to Harvard, Yale, Princeton, the California Institute of Technology, and a handful of other top American colleges.

"'The sacrifice would be worth it,' they said," Amanda repeated, almost to herself.

"What do you think?"

"I don't know," Amanda said, slowly. "We'll see."

Amanda had returned from the United States buzzing with doubts about her own educational experience, and I'd landed in China with just as many questions about our son's place in the system.

"Let's figure it out together," I told her, and she nodded.

Where Amanda was uncertain, Darcy knew very firmly where he stood on competition. "Competition motivates me. Rankings motivate me," Darcy told me.

Darcy was a tracksuit-clad teenager of seventeen when I first met him in Shanghai, a boy with smooth skin and a shy smile book-ended by dimples. A teacher friend put us in touch around the same time I'd met Amanda, and we found that our interests aligned perfectly. I wanted to interview a Shanghai high school student, and Darcy wanted to befriend a foreigner.

We began meeting every two weeks.

Where Amanda was the odd girl out, Darcy seemed to have everything neatly in place: He liked to tease his thick, black hair forward, so that it sat, frozen, like an ocean wave about to break on his sloping forehead. He was of good height. He dressed out of the pages of a teen sports magazine, with Nikes on his feet and latest model mobile phone in hand. He'd chosen the English name Darcy because he was drawn to the aloof yet debonair quality of Jane Austen's character. If Darcy's mother and father were to write a marriage advertisement to post in Shanghai's People's Square, where many anxious parents gather to match-make and try to marry off their children, it might say: "175 cm, born in 1996, a top Shanghai high school graduate, good moral standing, good qualities, destined for a top-tier university."

Darcy's only flaw, in his words, was his physique. "A body wasted away by study," he liked to joke.

At one of our early meetings, I was intent on drilling down the meaning of competition. "Competition might motivate you, but what about the shame that happens when a student finds out he's not at the very top?" I asked.

"When my scores get lower, a ranking gets lower, it motivates me to think about how to achieve more," he told me. "It pushes me to action." Once his math teacher complained that his former students were far superior. Darcy, sitting in the classroom that day, told me he felt ashamed. The next week he scored No. 1 on his math final.

"I wanted to prove my worth to him," he told me.

"But you are talking about external measures of worth," I told him. "Isn't there a downside?"

"Sure," he said. "Low-ranking students might eventually give up. But for those who are ambitious, rankings can give you a short-term goal to shoot for."

I looked over at him as he stirred his coffee. Like Amanda, he liked a strong cup of Americano. Unlike Amanda, he had few questions about his path.

"The Chinese education system is not perfect," he said, as if anticipating my skepticism. "It's a growing tree, and right now they can only focus on making the trunk strong. If the trunk grows well, then the flowers can blossom."

Part of making the trunk strong, I knew, was to continue this practice of filtering out students in a country of 1.4 billion people: Not everyone could make it through. Ranking and sifting the masses is a practice that dated back to dynastic times, when entire towns might gather in the main square or marketplace to await final results of the imperial exam. Back then, it was the *Zhuangyuan Ban*—the Champion List—that drew the crowds, just as Big Board drew the attention of parents and grandparents today.

The Champion List changed the fates of families, almost overnight.

⋆ ⋆ ⋆

I DECIDED TO read up on student life in imperial China. What was the day-to-day like for those kids whose sights were set on that Champion List? It turns out it was no easy road—then or now.

If you were a Chinese boy in the year AD 605, you spent most of your waking hours huddled over classic texts. Your future depended on how well you could memorize passages such as this one from the I Ching:

> When flowing water . . . meets with obstacles on its path, a blockage in its journey, it pauses. It increases in volume and strength, filling up in front of the obstacle and eventually spilling past it. . . .

The words were flowery, the meanings arcane. And there were a lot of them—more than four hundred thousand characters in total to commit to memory in the poems, speeches, passages, and annotations of the *Four Books* and *Five Classics of Confucian Thought*.

Daunted, you might feel compelled to drop your calligraphy brush and abandon study—one scholar estimated a young child would spend at least six years or more in seven-days-a-week study—but the emperor of the Sui dynasty needed ministers and officials. He'd decided a grueling, three-day exam would ensure the quality of men who would govern in his imperial court, and so an entire nation of male aspirants bent their heads over books in diligent study.

Girls, of course, needn't apply: Women with ambition could only educate themselves in hopes of supporting a spouse or son in their testing quest.

The ladder to social and economic status had many rungs, and on each rung sat an exam. First were the provincial-level tests. Then came the district tests, the metropolitan exams, and finally the national-level exams. The most extraordinary men might find themselves entering imperial court for the mother of all tests: *jinshi*. Passing the *jinshi* meant the possibility of catapulting to the top and snagging an imperial bureaucratic post in the country's capital. The emperor himself might look over your essay.

You cared about ascending this ladder, because to succeed meant a prize that was life-altering. An imperial position brought nothing less than prestige, status, and wealth for you, your family (and ancestors already long buried). Each rung afforded more status, and as you climbed, you distanced yourself from the carnage piled up at the base of the ladder.

And that carnage was thick, as pass rates were impossibly low. At the first rung, the district level, only one or two of every hundred candidates advanced to the next level, and the odds worsened from there. As each dynasty morphed into the next, passing the exams got less and less competitive, but as recently as the year 1850, still only one man in six thousand succeeded at all stages of the selection process.

The process was mentally trying. Painter Pu Songling of the seventeenth century sketched out the emotions of an exam candidate. The man was a "bare-footed beggar" when he entered the exam hall, and a "sick bird out of the cage" when he finished. After he failed to advance, he became a "poisoned fly" who burns all his books and thinks to "abandon the world." That "poisoned fly" might later decide to continue study, but this path required Olympic-level steadfastness: During certain dynasties the exam was offered as rarely as once every three years. It was a long time to wait.

This system was lauded for being a meritocracy (though, in truth, schools weren't universal and typically only nobles and merchants could afford tutors and books). In theory at least, both the peasant who can afford only rice for dinner and the noble man with turtle on his dinner plate could take the same test and advance through study, rung by rung. "Archaic, laborious, and daunting," researcher Justin Crozier called the imperial exam system, but it was also a "remarkable attempt to create an aristocracy of learning."

Today, just as in imperial China, that ladder theoretically drops from the sky into every corner of the far reaches of China. The message of hope is clear: Any child can hook a thumb onto a rung of the ladder, from the daughter of a nomadic shepherd in Xinjiang province, to a son of Beijing or Shanghai's urban elite. The student need only study hard, pass tests, and hope to advance to each successive level.

Of course, girls may participate now, and the system has experienced a number of changes, some brought about by emperors and others prompted by war or revolution. Yet the basic system of high-stakes testing and advancement remains: A child walks that same, narrow road for success that he always has.

For most Chinese today, particularly the lower and middle classes, a college degree is still the surest way to secure a stable job that provides upward mobility and employs the head, not the hands: Teacher. Doctor. Government worker. Marketer.

Yet the ladder to that degree is steep, with many rungs, and advancement still requires years of backbreaking, eye-straining study. At the first rung, five- and six-year-olds sit for entrance exams and interviews at the best urban primary schools. Then it's entrance testing for middle schools. From there things get really competitive with the National High School Entrance Exam, or *zhongkao*. Near the top of the ladder sits the vaunted National College Entrance Exam, or *gaokao*, which each year decides the fate of nearly ten million students and determines what college he'd attend and what field he'd study.

At these last two rungs of the ladder, the falloff is steep; between sixteen and eighteen million students take the high school entrance exam each year, and fewer than eight million go on to an academic high school (the cleanest path into a four-year university).

Each year, a population the size of London's will sit for the college entrance exam. Yet, only two-thirds of the nearly ten million teens who take the *gaokao* will get a university spot at one of China's twelve hundred colleges, and only 3 or 4 percent will ascend to what's called the top tier of colleges. The biggest losers on the ladder—roughly two million teenagers—are often relegated to non-skilled work or years of scrabble through self-employment or entrepreneurism in an increasingly competitive and resource-constrained economy. That's a lot of carnage at the base of the ladder.

Every June, the *gaokao* inspires a barrage of media photos, which indicate the fiercest of academic pressure cookers: Hangar-size warehouses with row upon row of black-haired heads bowed over exam papers. Students hooked up to IV drips for energy during test prep. Busloads of students on their way to exam sites, revving past thousands of pedestrians who raise arms in salute. Throngs of anxious parents camped outside the exam hall gates.

My favorite was taken by a photographer with a bird's-eye view of an entrance gate: On one side of the photo are hordes of students walking shoulder to shoulder to exam rooms, while on the other side, parents wait with sun umbrellas, bottles of water, snacks, and spare pencils. The children have been sent off to battle.

I once overheard a phone conversation between our ayi at the time and her thirteen-year-old daughter, a student back home in Anhui province. Though entrance exams are certainly the most critical, a relentless churn of weekly and monthly exams riddles the path of every

Chinese schoolchild starting in first grade. I could hear only Ayi's side of the conversation. In this and nearly every future call I heard, the topic was the same.

"Did you test already?" Ayi asked, speaking urgently into the mouthpiece.

"Which day did you test? The score didn't come out yet? Well, then, what did you think about how you tested?"

"Did it feel easy?"

"How many do you think you missed?"

"Then," she said, leaning into the phone, "how many do you think you *got right*?"

Because families have peered skyward at the rungs on those ladders for years, the need to study, and study hard, has always been ingrained in Chinese culture, even from a very young age—a hard work ethic becomes muscle memory.

I knew we'd probably relocate home to the United States long before Rainey would be of *gaokao* age. Yet, as long as Rainey was a child in this system, working alongside students trapped in this cycle of test-study-advance, study would have to be part of his daily routine. As Rainey grew older, Big Board's tally would change from height and weight to exam results and subject proficiency. This was inevitable; it was a deeply rooted feature of Chinese education.

For the moment, with what I knew, I felt okay with this prospect for my son. Growing up inside the competitive environment my parents created left me with a persistent, nagging feeling that other drivers were sprinting past me on the highway, though my own car was already at the speed limit. Yet there were also clear benefits: Like Darcy, I felt that knowing where I stood helped motivate me in key areas of my life and defined a clear goal to shoot for.

Still, I wondered whether competition set inside China's test-heavy system might deliver more detrimental effects, and I made a note to watch Rainey closely.

★ ★ ★

ON A WEDNESDAY toward the end of the school year, Soong Qing Ling School held its annual sports competition. The meet would pit the classes of Small Class grade against each other in a series of physical challenges and tests.

Teacher Chen, of course, had decided Small Class No. 4 must triumph.

A few weeks prior, Chen had sent out notices via WeChat, urging children to get plenty of rest and to rehearse "athleticism." "We have been practicing at school for this great challenge, and we urge you to prepare at home for the big day," Teacher Chen tapped out on WeChat. The parent cacophony chimed its affirmation.

Challenges might include a timed ladder climb, a sock-hop, or a tricycle race around Big Green. It all sounded innocuous enough, but I figured the character of such competitions would depend on the spirit of the leadership. Sure enough, the Great Ball Bounce of the following year would prove intense; in the weeks leading up to that meet, teachers handpicked parent-and-child teams and required them to huddle over basketballs at home, recording the number of bounces on gridded sheets. During morning drop-off, the school would set up practice stations on Big Green, and I observed hunched pairs of parents and children slapping down basketballs as quickly as they could. Lin Guanyu and his mother were among the victors that year, together logging an astounding 128 bounces in ten seconds. "We'd been practicing since March and expected a top-place finish! New records were created!" the teachers would effuse.

This year, the central challenge would be a game called Mountain Hole.

"We're most worried about this last game," Teacher Chen barked at a group of parents assembled in Small Class No. 4's classroom on the big day. The black bits between her teeth seemed as ominous as her intensity, and Rob and I joined the crowd, seated in child-size chairs arranged stadium-style, murmuring dutifully: Parent becomes student.

Chen laid out the competitive landscape. "There are five classes in the Small Class Grade Level, and two of them are fast—*lightning fast*. I've observed them during practices," Chen said. "For these two classes, the teacher might already signal her team has finished, and the other groups hadn't even started crawling yet!"

A collective gasp rippled across the room, and I saw that Teacher Chen was pleased with our reaction.

"We need to plan for action," Teacher Chen concluded. She described the Mountain Hole challenge in detail. Fathers would stand shoulder to shoulder on Big Green, rubber mats forming a path at their feet, with the children lining up at one end of the row of dads. From there, the race would unfold in two parts.

In part one, the first father would grasp the child at the front of the line and pass him to the next father, then repeat, until the entire group of children had been transported from one side of the father line to the other. Last child safely delivered, the men would then bend at the waist, hands outstretched to touch the other side of the mat to form a tunnel with their bodies.

"This tunnel is the mountain hole," Chen said. "For this part, the children must crawl as fast as they can in the direction from which they started."

The class whose children got down the line and back through the tunnel fastest would win. Teachers and classroom ayis would be placed strategically: one at the start to launch the children, another at the midpoint to urge for speed, and one at the finish to coax the kids home.

"Mothers? You'll be cheerleaders," Chen said, as she distributed pom-poms and horns, and I received a plastic contraption that wailed like a trussed-up duck. Battle instructions complete, we parents journeyed down to Big Green, descending the stairs in double-line formation. I clutched my horn, while Rob dutifully stepped right-left-right, mirroring the father in front of him.

We arrived to find parents from Small Classes 2 and 3 already in the heat of practice. Spurred on, our fathers stepped right into the action.

One man began passing an imaginary child to an imaginary waiting parent, twisting forward and back, forward and back like a weather vane caught up in a tornado. Another man began bouncing, knees bending and straightening, in anticipation of bearing the weight of a child. Rob stood off to the side, taking it all in.

"Shouldn't you be stretching or bouncing or something?" I teased my husband, as he snorted. Rob and I didn't relate much to the urge to win we saw in the faces of parents all around us, but we wanted to show support for our son.

When the children arrived, Rainey beamed in our direction from his place in line. Chen immediately began barking orders at the kids, a sudden urgency to her voice as other classes began assembling around us on Big Green. "*Stiffen* your bodies and hold your elbows *tight* to your waist as the fathers pass you down the line," she yelled.

For the parents, she had explicit instructions on gunning for speed: "Man One, grab a child by the elbows, Man Two grab at the waist. Man Three would alternate back to the elbows. Alternate the anchor point while passing the children. This way your hands won't get entangled," Chen said. "Maximum efficiency—and don't drop a child! The team will be docked five seconds."

"Five seconds!" a man bellowed, and the other parents murmured, pondering the size of this penalty. Chen offered more pointers for the tunnel part: "Men, suck in your bellies as you bend over, to allow room for children to crawl."

"I'm too fat!" one father yelled back.

"Then do a back bend," another responded, a needed break to the intensity.

"Are you ready?" the Soong Qing Ling principal suddenly proclaimed, from her perch at the head of Big Green, surveying the nearly 250 children and parents gathered before her, shirts color-coordinated by class.

The first child in our line wore red shoes, and he held his hands high above his head, mummified, waiting to be grasped at the waist.

The fathers crouched, rubbing their hands together in an attempt to roust up energy.

"*Yibeiqi*—Begin!" urged the principal. Red Shoes leaped into the first father's arms, and with that, the race was on. Instantaneously, the order of Big Green gave way to a screaming, maniacal mass of competitors. Parents yelled encouragement, children bounced impatiently, and teachers raced the length of their lines, urging speed.

Red Shoes was passed to the next parent, and then the next, and then on down the line. The next kid followed, and the next, and pretty soon each father was grappling with a kid, handing him off, only to quickly turn to receive the next child.

"Small Class Number Four—*Jiayou! Jiayou!* Add oil! Add oil!" Teacher Chen yelled, her heels rarely touching earth as she scrambled the length of the line.

One by one, the children were snatched and passed. Instructions were forgotten in the heat of the race. Most children forgot to keep their hands by their sides, and in their haste the fathers forgot to alternate elbows and waists. Instead, they began grabbing at whatever body part they could make contact with: underneath armpits, on the hips, by the upper arms.

"Add oil! Add oil! Add oil!" Children were grabbed and passed, grabbed and passed.

"Faster faster faster! FASTER!" Teachers Chen and Cai traversed the length of the line, clapping and chanting.

Red Shoes reached the end of the line, and he stood waiting at the west end of the tunnel. As the last child touched down, the fathers collapsed, hands touching down on the other side of the mat with their feet anchored in place, butts up in the air to form a tunnel. I recognized the back of Rob's blue jeans against the pants of the father next to him. The teachers elevated the chants as the children began to crawl. "Add oil! Add oil! Fast, fast, fast, fast, fast!"

The noise was astounding. The kids had also begun screaming, and together they formed a mass of moving arms and legs as they crawled, one child's head so close to the preceding child's rear that they almost looked attached. Like a chain of caterpillars, the kids shimmied on hands and knees toward the west end of the mountain hole.

From my vantage point, the tunnel's exit was obscured by a crowd of waiting parents gathered three and four deep. I noodled my way in, until I finally caught a glimpse of the end of the mountain hole, where the children were emerging headfirst.

There, I saw Cai and an assistant teacher in furtive heat. They were yanking children out and tossing them into a heap at the end. The children, blinking and dazed, got up and went over to sit on the curb.

They're pulling the kids out—that's cheating! I thought, in amazement. I stood on my tippy toes, peering over the heads of the thronging parents, but I couldn't see the western ends of the tunnels of the other classes. Was *everyone* cheating?

Teacher Cai began clearing the kids away from the end of the tunnel to prevent a traffic jam. The fathers chanted, "Faster, faster, faster!" I crouched down to peer at Rob, obediently perched in tunnel form about halfway down the line.

Abruptly, I saw Cai stand tall like a gazelle, craning her head to survey the landscape. There, she paused, decided it was safe, and pulled out a new tactic. Dropping to a crouch, she scooted into the tunnel to meet the next child head-on. She grasped the little boy underneath the armpits and scuttled backward, dragging him with her. Once out, she turned around and tossed him onto the grass. She did this a few more times, each time meeting the next child head-on in the mountain hole.

"Go, go, go! Add oil! Add oil!" the chants continued. I looked around, scanning faces. None of the parents standing with me seemed to register the fact that the leaders of Small Class No. 4 were unabashedly flouting the rules.

Finally, the last child passed the perimeter. Chen's hand went up fast as lightning—as fast as the lightning strike she'd ascribed to the other teams in practice—and where seconds before she'd aimed to escape attention, now she sought out the eyes of the principal.

"We won, we won, we won!" Chen announced loudly, her other hand joining the first in a frantic double-handed wave. I glanced at Teacher Cai, and she noted my look. I detected a shimmer of embarrassment.

The rest of the parents beamed, energized from the win, as the tunnel of fathers suddenly righted itself and began peeling away one by one, in search of progeny. I glanced over at the other races, but they were still in progress. Small Class No. 4 had cheated its way to victory!

My eyes found Rob's in the horde, and without a word we raised our eyebrows, almost in unison. I didn't want to be a killjoy, but I struggled to digest what I'd just seen. Children served an important purpose in Chinese society: making the adults in their lives look good. Here, the students had performed perfectly for Teachers Chen and Cai, who would now be celebrated by the principal before the entire school. An additional thought flashed in my mind: I knew that cheating was a well-documented problem inside the Chinese education system, and here I'd witnessed evidence of teachers taking shortcuts in the face of competition. This was just a kindergarten sports meet, but didn't it also set an example for the children of Small Class No. 4? Did Chinese students learn this type of behavior from the men and women who stood at the front of their classrooms?

Soon enough, it was time for the award presentation.

The principal brandished a handwritten certificate, which declared the winner: *CHAMPIONS for Small Class No. 4 in the Second Annual Sports Meet.* "I am healthy because I do sports," the certificate read, and Teacher Chen grinned as she grasped the paper with both hands.

Later, Chen and Cai assembled their children for a group photo. They posed under a red banner that stretched from tree to post and read SECOND ANNUAL PHYSICAL HEALTH COMPETITION.

In the photo, which I later slotted into a pile of Soong Qing Ling notices at home, the members of Small Class No. 4 are lined up in three rows, and they are smiling. The broadest grins of all belonged to Teachers Chen and Cai, who stand in the back, leaning over their tiny students like two Chinese elm trees. The children seem neither overjoyed nor burdened, but simply indifferent, which seemed to me appropriate expressions for pawns in a championship game.

I peered closer at Chen and Cai.

Both teachers held up two fingers in a V for victory.

PART II

CHANGE

6

THE HIGH PRICE OF TESTS

Learning something and taking a test on something are two completely different things.

—XU JIANDONG, A TEACHER AND FORMER EDUCATION OFFICIAL IN CHANGZHOU

Summer after Small Class year, we headed home for a break.

For Americans who live abroad, summer back home ushers in fresh air and state parks scattered throughout the United States, bookended by visits with friends and family. We hopscotched from Southern California beaches to Minnesota's pristine lakes to the humid, sweltering heat of Texas.

Rainey turned to me halfway through our trip. Liberated from his Mandarin tutor, blessed with lush green American parks and seemingly football field–size swimming pools, drowning in ice cream and hot dogs, my son had a simple question.

"Mom, why don't we live in America?"

While summertime kindled Rainey's curiosity about his parents' lifestyle choices, it gave me the opportunity to observe my little boy against American priorities and habits. I liked what I saw. In short, our little boy was winning accolades from American friends and family, against whose children Rainey's habits stood in stark contrast.

Rainey bounded down to breakfast with greetings for all the elders, which particularly pleased my father. He waited his turn. One time Rainey stood so patiently in a winding line at the American Museum of Natural History in New York that a stranger marveled, "How old is he? He's so well behaved."

His eating habits were decidedly un-Western, and his preferences came into clear focus against the American palate's mess of machine-processed nutrition. One friend's little girl ate only things that are white, such as pasta, rice, and bread. Rainey's cousin preferred food that came out of a box or plastic freezer bag: Chicken and fish nuggets were a top choice. One Saturday in Houston's museum district, I watched a little girl at the next table devour lunch. She stabbed a Capri Sun drink pack with a straw and ripped back the cover of a plastic pack of bread sticks, partnered by a container of chocolate goo. Dessert awaited her in a bag of bright orange Goldfish.

Soong Qing Ling school has an in-house nutritionist, and a chef whose sole job is to slap-and-pull fresh dough by hand, coaxing out delicate, elastic Chinese noodles. Green leafy vegetables accompany every Chinese meal. Rainey's school environment, along with Teacher Chen's take-no-prisoners approach last year, meant our little boy now eats eggs of his own free will. He happily eats whole fish with skin—nugget form not required—sheets of seaweed, and raw almonds, and he doesn't shy away from anything green, yellow, or orange.

One summer night, Rainey had sat down at the dinner table and reached for a piece of stir-fried bok choy. This surprised even me, and I put down my fork to stare at him.

"Watch!" Rainey exclaimed, stabbing a stem with a fork. He placed it in his mouth, chewed, swallowed, and met my stare triumphantly.

"Who taught you to do that?" I asked.

"No one," he said, reaching for another piece. "Just me."

Rainey relished the moment. Sometimes a kid must do things that he dislikes but that are good for him, and he enjoyed demonstrating his grasp of this concept.

"We'll tell Daddy," Rainey told me, plant matter moving down his esophagus.

Meanwhile, as Americans were rolling through a rebellion against testing—a New York teacher had likened the latest national education standards to child abuse—the Chinese school system was already preparing Rainey for an academic life. Now nearly five years old, Rainey had begun teaching his baby brother Mandarin—Landon was born two years into our time in China—two small heads huddled over a picture book, naming animals.

Only when I journeyed back to America did I glimpse the potential payoff of our education in China. As school had come to a close the previous year, Rainey had also begun to rejoice in his newest accomplishments in the classroom.

"Mom, I'm writing!" he had exclaimed proudly, showing me a worksheet he'd brought home from class entitled "Using the Telephone." Rainey had scrawled numbers in tentative, halting script. The "1" had been retraced several times, the stem of the "4" had been etched before the crook, and no line was straight, but there it was: my little boy writing numbers all the way up to 9. At the top of the paper, the teacher had drawn a fat star in bright red marker.

This development would likely trouble most American and European child development experts. Three and four years old would be considered too early to write, and a teacher's red star might also draw ire. Red pen is "like shouting," wrote several academics out of a University of Colorado study, and can "upset students and weaken teacher-student relations and perhaps learning." Western researchers suggested teachers use a pen of a neutral color. I'm certain Teacher Chen would snicker at the idea of American universities spending time and money researching the emotional effects of a teacher's pen color; personally, I was thrilled that my child—who still wore Pampers at night—was etching out numbers and earning star marks, in flaming red.

That's not to say Rob and I didn't have concerns.

Close friends in Los Angeles had placed their toddler in a school that bills itself as "humanistic, experiential, play-based . . . while [supporting children] with the deepest respect for their dignity and self-worth." At the school, aptly called Play Mountain Place, children design their own curriculum, and there are no grades, punishments, or rewards. Children may skip lunch if they want, or wear diapers until they're six (no potty training required!)—the choice is theirs. Even in the primary school years, Play Mountain Place students aren't required to learn to read or write or do arithmetic. Teachers treat children like adults capable of making their own decisions, though certainly, chaos sometimes rules. "Imagine *Lord of the Flies*," our friends Jen and Kevin told us. "But the school's philosophy is that academic success is only one way to be a human being, and that children should develop in their own skin, at their own pace."

It's fair to say that when I thought about Play Mountain Place, when I wasn't chuckling at the school's audacity, I was paralyzed with insecurity about my decisions for Rainey. The school was an extreme choice by any measure, but still, I wondered whether it might pump out confident, natural leaders who could independently direct playdates as well as their own academic pursuits. Conversely, would the Chinese way only produce mathematics mega-nerds who would ultimately end up—gulp—working for the adult versions of those Play Mountain Place kids?

The reality would never be so black and white, but my fears were always glaring and stark.

So, I concocted a master plan to offset what I considered deficiencies in Rainey's Chinese schooling, and home would be our domain. Safety was nonnegotiable, as it is for the Chinese, but for almost everything else, Rob and I encouraged our son to decide for himself. We populated his room with markers and art supplies, and infused his environment with choice. Rainey could make decisions such as what he'd wear, what books he'd read, and what sports he'd play. Musical instruments were strictly optional. On low-pollution days, we headed to the soccer field or the tennis courts and spent hours kicking a ball or rallying over the net. The Shanghai government had also put muscle behind developing a thriving cultural scene, and to expose our kids to art, we began frequenting some of the city's museums, which were beginning to bring in international exhibitions. We also pursued activities that had no purpose other than leisure, such as fishing, which is purely inefficient from the Chinese point of view. "Why don't you just buy fish at the market?" a Chinese acquaintance once asked me.

Plan in place, we readied ourselves for another year of Chinese school.

On the first day of school of the new Middle Class year, Rainey bounded out our front door.

"Let's go, Mom!" Rainey squealed, as we rode the elevators down to street level.

Rainey dived headfirst into the mayhem of the streets, skipping a few steps ahead of me. "Slow down!" I yelled after him.

The street vendors were out. A weathered man from Xinjiang balanced a bamboo rod across his shoulders, a basket of cherries on one end and a platter of purplish *shanya* fruit on the other. If he walked straight, rod and all, he'd clear an eight-foot-wide section of sidewalk. Instead, he sidestepped gingerly, winding through the sidewalk crowd until he found a place to stop and display his wares. Bicycles and electric scooters weaved in and out of a glut of vehicles, which had slowed the pace of traffic on the street to that of a parking lot. Vehicle exhaust drifted into my nose.

Rainey continued to bound ahead of me, past the bustling hospital and eventually into the alleyway that led to Soong Qing Ling. We passed a pigeon vendor and his squawking birds, and entered a maze of *shikumen* lane homes with laundry flapping from windows. Down this alley, a little vegetable market was under way, with wizened vendors unfolding canvas tarps on the pavement. The tarps cradled heads of white cabbage, green bok choy, garlic, chives, and plump orange carrots.

"How do you sell these?" a customer asked, tapping a fat, cream-colored daikon.

"Twelve kuai a kilogram!"

Strolling through a Chinese neighborhood accosted all the senses.

I was glad to be back, and so was Rainey. Yet, as we approached the familiar iron gates of the school, Rainey suddenly slowed, as if anticipating the transition from an American summer to the Chinese schoolroom would be jarring.

"Wait, Mom, wait," Rainey said, slowing. "I don't want to go. The teachers are going to say that they're going to call Mommy if I'm bad, but then they don't call."

I caught up alongside him. "Honey, you have a new teacher this year, and new classmates. Let's see how the day goes, shall we?"

The first month at Soong Qing Ling, a new element of order emerged in the morning routine: Blaring Bullhorn at the entrance gate. Principal Two had begun standing just inside the gate, holding a loudspeaker larger than her head, and she took to screaming at parents and children as if they were cattle about to miss the closing gate.

"Don't be late, don't be late," she blared, as parents and grandparents scurried. "We will close the doors at nine a.m. If you are late you won't be able to come into school."

We were called out the second morning, even though I'd tried to keep my head down as I entered the gates.

"Rainey, it's 9:02 a.m.," Blaring Bullhorn pronounced, as my little boy's name echoed across Big Green. "If you're late tomorrow, you cannot come into school."

"*Buhaoyisi*—we are very embarrassed," I told Principal Two's bullhorn, since I couldn't see her face. I nearly kowtowed. Rainey shrank against my legs, and we scampered up to the classrooms.

From the outside, Soong Qing Ling's discipline was as fierce as ever. Inside, we would discover, things were changing.

* * *

"WE ARE TRYING to adopt some Western ways."

Rainey's Teacher Chen had uttered these words during our parent-teacher conference the previous year, offering her defense as I'd inquired about threat making in the classroom.

As I stared at Rainey's new master teacher for the Middle Class year, the context for that comment began to materialize. "Affection in education is important," Teacher Song told a group of parents in a fourth-floor conference room.

"Adults should learn to express their love for children, so they become loving people full of affection," Teacher Song told us, from her perch at the front of a rectangular table carved of rosewood. Ballerina slender, she had black hair that shone sapphire-blue, and a mole on her chin that seemed to dance as she talked. Where last year's Teacher Chen had dark teeth that seemed ominous, Song had only gleaming bright whites, which I noticed because she liked to smile. Already, this was an immediate sea change from last year.

"We don't force young children to do exercises," Song continued. "Rather, we incorporate learning naturally into our daily activities. We learn the basics gradually."

Was I hearing correctly? Who was this person, and how long would she last at Soong Qing Ling with such radical ideas?

"Painting is not about strictly following the teacher's example," Song continued. "We expect the children to find a way to express themselves through signs, lines, and shapes. We teach basic painting methods, and leave a lot of latitude for them to explore."'

I found myself nodding, but Song offered counsel for parents who might find this kind of exploration unruly or even offensive. "We shouldn't judge a picture by whether it looks similar to a model," she told us. "Don't tell your children 'Your painting is so ugly.' Just encourage and compliment the kids—they will perform better when they have confidence."

It was a classic argument for self-esteem.

Song even had decidedly un-Chinese advice about prepping for an academic career. "Don't jam knowledge down your children's throats," she said. "Middle Class children need a little bit of pressure but not too much."

I glanced around the table. Some parents nodded, while others wore faces of doubt. I took a closer look at the skeptics. Sitting at this rosewood shrine to education was a mom who'd worked with her child to memorize one thousand Chinese characters over the summer. Others had commenced piano and flute lessons for their kids. One parent had snapped a picture of his son's cram school math exercises and sent out proof of the boy's diligence over the WeChat parent group.

Song had words for this kind of parenting. "My daughter is in the fourth grade now, and I never taught her anything when she was in kindergarten," she said. "Let your children enjoy their childhood." She made sure to scorn the top-scoring child in her daughter's class as an over-tutored and highly scheduled kid. "A kid like this will burn out before he enters the workforce," Song told us.

Frankly, Song sounded a bit like a teacher from Brooklyn or Los Angeles, pushing a progressive education and urging us to opt out of the toddler rat race. I was encouraged by her words, but wondered whether her philosophy would translate to action in the classroom. Time would tell.

Song concluded our meeting with some distinctly Chinese goals, which had to do with eating.

Children would train their "chopstick muscles" in class this year, she said. They'd learn to debone butterfish and peel the shells off shrimp without assistance. And, as always, the pace of eating is important, Sun said, because "winter is coming."

"Food cools very fast in winter, so we've started training. Lunch must be finished in thirty minutes or less; otherwise it gets too cold," Song explained. At this, I laughed. The Chinese have strict rules about the temperature of anything they swallow, and they felt strongly enough to incorporate guidelines into the educational curriculum. Food must be eaten hot to maximize digestion and absorption, water is never to be drunk cold, and stir-fry foods are to be chased by hot tea or warm water to melt the oils in the tummy. (Order a beer at a restaurant in China, and the waitress will ask you whether you want it warm.)

Overall, much of Song's approach sounded like second nature to me, things I might do as a matter of instinct: Talk with kids about topics that interest them. Listen patiently. Read books together. Encourage children to work independently, and enhance their self-esteem and confidence.

This was entirely contrary to what I'd experienced so far in Chinese education.

★ ★ ★

WHAT EXPLAINED TEACHER SONG?

A Shanghai kindergarten principal offered a clue one morning. I'd requested an interview to talk about trends in education, and in response she handed me an official-looking booklet. Issued in the 2000s by the Ministry of Education, it was titled *Guideline of Children's Learning and Development for 3- to 6-Year-Old Children*.

I glanced down at the book, with its white cover and orange-and-black print. I gave it the nickname White Bible, a fitting moniker, I thought, for the government's attempt at steering early education in a kinder, gentler direction.

"This is a very important book," the woman had told me, patting White Bible's cover. "All kindergartens in China must abide by this. It says that everything should be based on children's natural rate of development." Intrigued, I flipped through the booklet to find dozens of proclamations, all designed to protect the sanctity and individuality of the Chinese child.

"Every child will progress at different rates," White Bible proclaimed. "Individual differences should be respected." "Do not use 'one ruler' to measure all children."

In other words, I thought, don't measure children's height, weight, and recorder prowess and post rankings in a public corridor?

"Maintain positive, happy emotions for children," proclaimed White Bible.

Meaning, don't threaten children with police capture, as Rainey's teachers had last year?

"Encourage children to develop, with the courage to explore and imagine," White Bible proclaimed.

In other words, Little Pumpkin should be allowed to draw rain in any color he wishes?

"Guide children through experience and hands-on learning rather than indoctrination and pursuit of knowledge," White Bible stated.

Meaning, don't cram children's heads full of facts and Chinese characters during summer test prep?

Then this: "Allow young children to make mistakes. Do not beat them—they will be so afraid of punishment they will tell a lie."

I blinked and looked again. Clearly, the Chinese government felt it needed to spell this one out.

Much of White Bible, I learned, was borrowed straight from the West, and it was required reading for kindergarten educators throughout China. With the help of UNICEF (United Nations Children's Fund), Chinese researchers had consulted early childhood development guides from thirteen countries, including the United States, United Kingdom, Germany, and France, to help educators better encourage the physical, mental, and moral health of young children.

I'd heard vaguely of attempts at system-wide reform, and when I finally delved into its specifics, I was surprised to find that China was looking to the West for lessons about nurturing the whole child, just as we'd been looking with envy at their academic achievers. In fact, China has been trying to make its school system friendlier and more welcoming for children of all ages, from the youngest toddlers all the way up to college students, and this effort had begun decades ago.

White Bible was just the latest aimed at kindergarten reform. Clearly, some educators took to its directives like second nature, while others—like Pumpkin's teachers—were battling long-held attitudes and behaviors, which kept their classrooms traditional and their approach authoritarian. I could name violations I'd seen or heard about, which infringed upon nearly every one of the book's tenets. Yet this booklet was evidence that, at least philosophically, China was trying to move in a friendlier direction in educating its children.

"There is nothing that should remain unchanged when it comes to reform of our educational institutions," Ministry of Education official Wang Feng told me during an education conference in Beijing.

What had instigated this kind of effort?

★ ★ ★

IN 2004, MA JIAJUE got very angry, and what happened next made him famous.

Ma was a scholarship student at Yunnan University, described by his family as studious and shy. He'd run away from home once during high school because he was afraid he'd fail the *gaokao*, but soon enough he came back around. These were fears familiar to any Chinese grade-schooler, and for the most part, Ma Jiajue was unremarkable in nearly every way.

Until the day a handful of college classmates accused him of cheating in a card game.

At that, Ma took a knife and promptly killed his accusers—four students, including his roommate—and stuffed their bodies into dormitory closets. Then he went on the run. "Student Killer an Introvert Who Finally Cracks," proclaimed *China Daily*. A nationwide manhunt ensued, and Ma was finally arrested in Sanya after twenty-one days.

Another famous case of social malfeasance came when a Tsinghua University student tossed sulfuric acid in the faces of several bears at the Beijing Zoo. One bear was blinded, the others injured. More recently, a handful of universities have documented cases of students poisoning classmates, sometimes resulting in death or paralysis. One incident involved the 2013 death of a student named Huang Yang, after his roommate laced the dormitory water cooler with toxins.

These events typically prompted a national soul-searching about education in China. Why would Ma Jiajue kill his classmates for a mere insult? What about his life had turned him a sociopath? Why would a student at one of China's top universities treat bears at a zoo so inhumanely? Why would a student poison his roommate?

These were extreme incidents, yes, but state bureaucrats and educators seized on them to continue an impassioned plea for change. "These events are a shortcoming of our education. It is our wake-up call—we can't focus only on students' academic development—we must also cultivate healthy personalities and good emotional competence," said Wei Yu, vice minister of education for nearly a decade. "Our education lays emphasis on competition and individual success. Chinese students are clever and hardworking but lack the spirit of sharing and cooperation," said a professor during an interview for *People's Daily* online.

China's leadership had intended to fashion a school system to create well-rounded, moral, athletic, hardworking contributors to society. Instead, what they got is a population of "test-taking machines" without social and emotional abilities, as many a media report put it.

It was clear where the blame should lie for creating generations of professional students: China's test-based education system, or *yingshi jiaoyu*, say experts and government officials alike. Government documents themselves condemn the system with harsh words: It's a radical departure from the basic needs of learners. It buries students under mountains of homework. It makes scores the only measure of worth of a child. It hurts student's initiative and creativity. It violates the original tenets of the Education Act itself. Amanda, my friend in Shanghai high school, told me, "I've never been allowed the time to learn how to deal with people. I know books, but I hate human contact."

The bodies of schoolchildren are wasting away under the weight of their books. The eye doctors of China would tell you that children have vision problems like never before; four out of ten elementary school kids are nearsighted, a doubling of myopia in China over the past decade. Middle school kids fare even worse. Genetics, perhaps, but the media and commentators seized on what they saw as proof that change was desperately needed. And while students' brains are growing fatter with knowledge, so are their bodies: One in five school-age girls is considered obese, and for boys it's one in three (although changing diets are partly to blame).

"China has paid a very high price for focusing on tests," Guan Yi, a teacher at Suzhou High School, told me. "Learning something and taking a test on something are two completely different things," said Xu Jiandong, a former government official in Jiangsu province.

China's highest education officials haven't sat by idly; in fact, for decades, they've tried to move the system away from testing. Let's usher in a more "holistic" approach to education, the ministry decreed in the 1980s, adopting a concept called *suzhi jiaoyu*, or "Quality Education." This was part of a government campaign to lift the quality of the entire Chinese population. (The one-child policy was part of this effort.)

This "quality education" umbrella effort gave rise to dozens of policies. When officials found that toddlers were learning content appropriate for much older children, they declared, "Let us *de-elementarize* preschool!" That became the title of a policy document. When research showed preschool games helped lighten the academic load for young children, "Games and Play" became the name of another. The decrees came fast and furious, and they addressed the education of preschoolers all the way through college. Altogether, hundreds of directives came out of the central ministry, with fancy names such as "Kinder, Gentler," and they attempted a variety of changes: They prohibited elementary school entrance exams; shortened the school day; slapped a time limit on homework for grade-schoolers; banned numerical grades; and prohibited makeup classes after school, on weekends, or during holidays, just to name a few.

All were aimed at lightening the academic pressure on China's compulsory student population—roughly two hundred million strong—and indeed "Lessen the Burden" became the title of another policy. Education bureaus at provincial and local levels attempted their own initiatives. Shanghai's education ministry, for example, recently required that middle and high schools employ full-time psychological counselors to help students cope with academic pressure.

Entrance exams are also getting a hard look, as the government attempts to diminish their importance for high school and college admissions. In Shanghai, the most recent reform considers calling for students to have *two* chances to pass parts of the *gaokao*—with the idea that the chance for a do-over makes the first time less of a nail-biter. In 2017, Shanghai universities will also consider outside-of-school activities such as volunteering, as well as the scores of regular, routine high school tests. I get the trend: More and more, universities are being asked to consider the *whole* student—instead of only that single *gaokao* score.

As I studied these reform efforts, which seemed to point one way and then reverse course to try something altogether different, one thing became achingly clear: The Chinese government's long-standing dissatisfaction with its school system, as well as a paralysis when it came to fixing it once and for all.

★ ★ ★

INTENTIONS TO CHANGE are well and good, but this parent has a horse in the game. I wanted to know about outcomes. What did these reform efforts mean for a student's daily life? What did they mean for Rainey?

I discovered that the Education Ministry and reform advocates had been taking the pulse of its patients—the country's school system and all those involved—in the form of surveys. Most recently, it surveyed nearly five thousand primary and middle school students, principals, and parents all over China. But the results, released in 2013 after several generations of reform had come to pass, were disheartening.

Only one in four students felt their workload had lessened. Students were still buried under homework. Most kids were still taking outside prep classes, despite an outright ban on extracurricular study.

More than three-quarters of students suffered from *yanxue*, or a hatred of study. The results hadn't changed much since 2005, when a wide-ranging survey found most students had no time for housework or exercise, and indicators of study fatigue dotted the report with words such as "depression," "boredom," and "anxiety." Today, one in three students say they actually have an even heavier workload than before.

Despite decades of work, educators had measured no significant progress in reducing academic pressure. The culprit was clear: All those tests were making it impossible for students to do anything other than bend their heads over books, and entrance exams for high school and college—still firmly in place—only intensified the pressure.

I spoke with a Shanghai psychologist who had been installed at a middle school as part of a reform effort. "My office is often empty. I'm trying to cure *yanxue*, a hatred of study, but the students are often too busy taking tests to come talk to me," said Yang Lingqiong.

The former education chief of Yunnan province, Luo Chongmin, gave a biting assessment in a speech. "Any policies are completely 'empty and unfeasible because *gaokao* is still there,'" he ranted. "The only way to really relieve the burden of study is to change the current evaluation system, which is exam-oriented. Middle schools and high schools don't dare slow down their pace, the Ministry of Education has no real power to punish schools, and officials and civil servants never carry out any supervision. Lessen the Burden has been going on for more than half a century, but the burden on students has in fact increased greatly."

Looking carefully, I saw that this burden was actually revealed in the same PISA (Program for International Student Assessment) test that placed Shanghai students tops in the world in math, reading, and science. That exam had also surveyed students' quality of life, creative thinking skills, and time spent in study, and in these areas, the Chinese saw only slashing red marks all over their test papers.

In fact, the man commonly dubbed the grandfather of Chinese education reform looked with jealousy across the straits to Taiwan, South

Korea, and Japan. "Those countries scored right after Shanghai—second, third, and fourth—but they spend a third to a half the time Shanghai students do in study," Yang told me in Beijing in 2015.

He's right: On the amount of time kids spent poring over books, Shanghai, too, was also the world's No. 1.

There's one more important thing: American kids may score lower on paper tests, but those who excel in math love the subject from their heart. "Chinese students have better scores but their interest in math is not sustainable," said Liu Jian, a mathematician who works on China's national curriculum. He cited research that shows Chinese students' interest in a subject is more externally driven, by exams, attention from teachers, rewards, and the announcement of scores. Once those external rewards are taken away, Liu told me, their interest and motivation to learn wane. "If a student is not interested in learning, it will be difficult for him to achieve great things in the future," as Beijing professor Gu Mingyuan put it.

Instead of celebrating their top-ranking perch, Chinese educators had questions: How did Finland achieve their results in half the homework time? And how can we get our children to love learning for the sake of learning itself?

★ ★ ★

HERE'S THE CONUNDRUM facing the Chinese Ministry of Education.

When entrance exams are less important, students would certainly feel the pressure lift away. But then, educators must come up with another way to select kids for high school and college. That would mean introducing such things as essays, interviews, and teacher references.

"That requires human graders, and they're prone to making all sorts of mistakes," said Andreas Schleicher, the architect of the PISA exam. "Once you introduce human graders in a high-stakes environment, you raise all sorts of questions. You can imagine what happens."

The system simply isn't ready for gray areas: China must process millions of students each year from thirty-four provinces and municipalities all over China, and filter them into one of nearly three thousand higher education institutions. And there aren't enough spots for everyone.

"So, for a hundred percent reliability," Schleicher concluded, "you always go back to a multiple-choice test." This is the black-and-white option, which theoretically leaves no room for corruption and mayhem. In principle, then, *gaokao* is as close to giving everyone a fair shake as you'll possibly get.

Chinese culture also prevents change; the country suffers from a societal loyalty to testing, which grants social status based on scores. A hard habit to break, all right. "China is a nation of test addicts," says educational consultant Jiang Xueqin. "Video games are addictive, and tests are addictive for the same reason. You do a math test, next day you find out you get a 99, 'I'm a great person, right?' It's an instant feedback loop."

I understood this sentiment well. More than fifty years after my father took the National College Entrance Exam in Taiwan, he can recite his total score and the individual tally for each of the six subjects of math, chemistry, physics, Chinese, English, and Chinese constitution. You can still see that pride in his seventy-year-old shoulders, as well as the joy of having been admitted to Taiwan's top university. Identification by number extends to the community; as an adult, I accompanied my father to his college reunion and marveled that, over dinner, he and his classmates could still cite each other's entrance exam scores and subsequent rankings in college four decades after the fact.

When it came my turn to sit for what you might call America's equivalent—the SAT (Scholastic Aptitude Test)—my father liked the result so much, he locked the score in his memory. Fifteen years later, as he toasted Rob and me to a lifetime of happiness at our wedding, he quoted the number before two hundred guests over filet mignon. "My daughter was a National Merit Finalist," he'd announced, as I shrank into my seat at the head table.

The message: That number was a measure of my worth as a human being.

Long after a person's formal education is complete, China still likes to test—as if the country itself were an addict. Qualifications exams exist for all sorts of professions including doctors, therapists, lawyers, and judges; and the civil service exam sits at the gateway to a government job. There's also a test for standard Mandarin speaking. "America doesn't have one for spoken English?" my Chinese friend Meredith asked, stunned when I shook my head. "In China it's very important." The standard for cops is a score of 80, for teachers it's 87, and for actresses it's 93. For some professions, the score is only a recommendation. For broadcasters like Meredith, whose voices radiate over national airwaves, a minimum score of 97 is mandatory.

"I got a 98," she said, proudly. "You're supposed to take it every five years. It's very important to have standard Mandarin."

I discovered something else, which turns out to be the Chinese school system's straitjacket. Tests in China are actually used to assess teachers and schools. Generally, teachers are evaluated on their students' test scores, and schools are ranked on measures such as how well they graduate students into the next level of schooling (which is often determined by entrance exams). As much as everyone hates testing, teachers need their kids to focus on it. (This was becoming more and more true in the United States, too.)

In fact, in China, a high school teacher's salary can be boosted up to 20 to 40 percent if his students' *gaokao* scores are high, estimated Wei Yong, a teacher at Beijing National Day School. Schools win and lose reputations based on their students' collective *gaokao* scores; high scores are a gift that keeps giving, as administrators can justify charging more for school incidentals, and it's not uncommon for government funding decisions to be based on performance. Results are collated and ranked by class, school, city- and province-wide, like a public scorecard.

"No one truly wants to reform," the former headmaster Kang Jian told me. "As long as *gaokao* is an evaluation for education, the pressure will always be there."

He added ominously, "And *gaokao* has only gotten more important."

* * *

THE CHINESE PROVERB *jin ri shi, jin ri bi* means "Don't put off for tomorrow what you could do today." Another Ming dynasty poem ponders if you "delay something until tomorrow, how many tomorrows can you have?"

Don't procrastinate, as the proverbs go, since there's always a task to be tackled or a chore to complete. This lesson is ingrained in my own muscle memory—as much as I've tried to fight it—and for every Chinese member of my immediate family, unfortunately, relaxation is a learned skill. Even today I struggle with guilt when I kick back with a glass of wine after a long day's work. Shouldn't I be rebalancing my retirement portfolio? Cleaning the kitchen? Exercising?

One day Teacher Song tapped into this sentiment over WeChat: "The school will host a distinguished panel who will speak about elementary school readiness and admissions," she wrote, the ultimate in *jin ri shi, jin ri bi*, given that elementary school was nearly two years down the road.

"Rainey's finally adjusted," I cried to Rob, wishing I could kick back and relax on this one. "Can't we coast for just a little bit?"

"We're probably already late in thinking about it," he replied, shaking his head.

As always, Teacher Song insisted on an immediate reply: "Please let me know if you'd like to reserve a seat." So many RSVPs flooded in that the school was compelled to plan for a second session.

I once had a glimpse at a test for one of Shanghai's top three elementary schools. Three pages long, it was full of problems written in Chinese characters, with multiple-choice answers. For example:

> There are two glasses of juice. The child drinks half a glass, and Mama pours it full again. The child drinks another half a glass,

and Mama pours it full again. Then the child drinks everything up. How many glasses of juice has the child had to drink?

This is intended for five-year-olds? I'd thought to myself.

"*Buhaoyisi*—let me bother you for a moment." I pulled Rainey's teacher aside one day. "Aren't primary school entrance exams supposed to be banned? I thought the system was trying to lighten the testing. Why should we prepare for them?"

"Yes, but prestigious schools still have entrance exams," Teacher Song said, glancing my way with surprise. "It's important, because some of the best high schools are part of a system, and acceptance to the right track starts with primary school." This was the same reasoning uttered by Gregory Yao, the extracurricular-obsessed father who'd put his daughter in "early MBA" classes as a toddler.

I looked at Teacher Song, processing my first inkling that her classroom might deviate from the White Bible words she'd championed in that first parent meeting. How could it, inside of this culture and system?

Song and I had developed an easy rapport. Teacher Chen, Rainey's teacher last year, saw a foreigner needing reeducation, but Song was more likely to see the nuance of my existence in China. "You look like 'us Chinese,'" she'd said to me once, quizzically, giving a nod to my Mandarin-with-a-foreign-accent.

"I am like 'you Chinese,' except I'm American," I'd assured her. "My parents are Chinese but emigrated to America when they were young."

"*Zhidaole*—I understand," Song had said, nodding her acceptance.

On Teacher Appreciation Day, I decided to deliver a Coach wristlet, determined to get our relationship off to a healthy start. Rainey traced out a dinosaur onto cardstock, and I slipped his handiwork inside. Teacher Song surveyed my offering—small enough to nestle in my palm, discreetly wrapped in white tissue—and decided to accept.

Overnight, she changed her mind. "This is expensive," she said the next morning, pulling the package from a hiding place inside a bin of markers. She quickly pressed it into my hand. "Sorry, I cannot take this."

"It's just a small token of our appreciation. I hand-carried it from America. No import taxes! It's not that expensive," I said. It was textbook "politeness strategy."

"Can you give it to your friend or your mother?" she said firmly, indicating that the conversation was over. For now.

She was playing her part in the "reoffer-decline exchange," and I nodded.

Even though she'd refused my gift, she noted my gesture, and according to the protocol of Chinese gift-giving, I was slowly working my way into the club.

Song's WeChat profile photo showed her sitting on a ledge with the ocean undulating behind her. A little girl perched on her lap; she looked to be about eleven years old. Like me, Teacher Song was the mother of a child in China, forced to navigate the system. We were both trying to find the best way forward.

On the day of the preparedness seminar, I filed into the assembly hall alongside 150 other parents. Onstage were three mothers whose children had graduated from Soong Qing Ling into higher levels of schooling. The mothers were dressed formally and seated before microphones, and it was the second speaker who resonated most with the audience of anxious parents.

"With the more and more flat world, our kids should not only have a good mastery of Chinese, math, and English but also art and philosophy," she told the group of parents. Mother of three children, the woman clearly had the money or connections needed to skirt the one-child policy, either by giving birth overseas or by paying fines.

"We all need a blueprint, a rough draft of your children's development plan. What's the potential secondary school after primary school? Or should they go abroad? Or is it better to finish bachelor's at Fudan, Jiaotong, or Peking University?" she asked, naming the country's elite universities, "then pursue master's degrees abroad?"

Some parents scribbled notes, while others trained video cameras on the "preparedness gurus."

"We're all on this path," the speaker was saying, "not only for our children but also our children's children. That's how important our decisions will be." She spoke of a television series from nineteen years ago called *Yongzheng Dynasty*. "I'll never forget when Emperor Yongzheng said, 'Don't merely choose the new emperor; also take into account the next heir, and the next one.'"

The parents—all with toddlers in the classrooms just down the hall—nodded as they thought of the futures of their unborn children and grandchildren. Murmuring agreement rippled through the crowd.

In this room, I thought, glancing around at the nodding heads of Soong Qing Ling parents, were 150 reasons reform would not be coming quickly to Chinese education. The forces of culture and a test-based system were too strong to ignore.

I would soon discover a new reason that change was slow to come. For an independent thinker raised inside a democracy, it was the most unsettling of all.

7

LITTLE SOLDIER

The school curriculum is being used as a "means of political indoctrination for the purpose of ruling the people, rather than for the development of the individual person."

—LI MAOSEN, MORAL EDUCATION SCHOLAR

One night, Rainey had an announcement for me.

"I like China better than America," he said, in his singsongy voice.

"What, Rainey?" I asked, startled. We'd never really talked about the concept of country, nor had we ever discussed nationality or ethnicity.

To explain, I would have to have said: "Rainey, my ancestors are from China, and your father's are from Germany, Norway, and Sweden. Your father is as white as they come, but in an odd twist, he lived in China as a Peace Corps volunteer long before your mother ever stepped foot in her ancestral motherland. Racially, you're exactly half of each of your dad and mom—Caucasian and Chinese—but your *nationality* is American. Oh, by the way, all your classmates are Chinese nationals."

It was complicated. I glanced over at Rainey and tendered a question. "Why do you like China better than America?" I asked.

Was Rainey repeating something he'd heard at school? Had someone planted this idea in his head? And what about that time back in the States, under a classic barbecue-and-summer blue sky, when America had won?

Rainey lifted his shoulders in a four-year-old shrug.

★ ★ ★

AT THE FRONT of most classrooms in China sits a framed flag of the People's Republic of China. Bright red to signify the Communist revolution, the Chinese flag boasts a large yellow star, as stand-in for the Communist Party, and four smaller stars arranged in orbit—the people under its rule.

The Chinese schoolroom flag is nothing like the American version I remember; the Stars and Stripes hangs from a pole within a child's reach, its fabric rippling whenever the classroom door sends a gust of air its way. You can touch it, wipe your grape-jam-smeared mouth on it—not that your teacher would be happy about this—or even burn it if you so desire, with your right to do so protected by the United States Constitution. By contrast, it's a crime to desecrate the Chinese flag, punishable by prison time. In the schoolroom, the Chinese flag is affixed behind a pane of glass, encased in wood, and hung high above the students' heads, a fitting metaphor for education in China: Always heed the higher authority. Mounted north of the chalkboard, and hovering over every schoolteacher in China, is this daily visual reminder of the true purpose of Chinese education.

Chinese schools teach math and science, yes, but they are also charged with a singular objective: shaping students into proper citizens of their country.

Much is required of the mind of the Chinese citizen. He must love country (China), his people (the Chinese), labor (*working* for China), scientific knowledge (the key to China's economic future), and socialism (the Chinese market model). These "five loves" are embedded in the national curriculum and appear in textbooks from primary school all the way through college, and they are the pillars of the Party's campaign to shape the worldview of the people.

It's patriotism as a governing technique, a practice as old as China's history itself. Indeed, you could say that China's leadership has had thousands of years of practice tightening its grip on the hearts and minds of its people. Confucius himself believed that government by ideology was "more important and effective than government by law," wrote moral education scholar Li Maosen. Confucius said: "He who rules by moral forces is like the pole-star, which remains in its place while the other stars surround it."

In other words, why use guns or force when leaders instead could cultivate an internal, self-governing compass in every person under your rule? I thought again about those families who stared upward at the Champion List; years of backbreaking study were required to advance into the imperial court. Yong Zhao, a scholar famously disdainful of China's education system, said the exam system served a convenient purpose for the leaders who perpetuated it: Keeping the best and brightest young men memorizing texts most hours of the day, "their minds . . . steeped in Confucian philosophy, which forbade them to have any unorthodox thoughts." Lots of time devoted to study meant little time to organize rebellions, making exams a handy tool for governing a large mass of land carved up into, at times, warring factions.

The Party's approach to education today is little different. In 1949, the Communists took the helm of a country ravaged by war, with vast gaps in welfare from village to village and region to region. It inherited a school system that was in tatters. Slotting patriotism into the school curriculum was a neat trick: It brought together a widely divergent population, since the schools the Party was charged with uniting were as disparate as the work-unit schools of Mao's Communists, the institutions of the departed Kuomintang, and classrooms inside orphanages governed by provinces.

In the early years of Communist rule, schoolteachers were directly responsible for disseminating Party policy. The first primary school textbook created under Mao sang the man's praises: "Chairman Mao is like the sun; he shines even brighter than the sun . . . we will follow you forever." A 1950s textbook passage reminded students whom to thank for progress:

> My grandpa herded sheep when he was six; my father fled famine when he was six. This year I turn six, and I am in school due to the Communist Party's help.

The state's propaganda department was nothing if not opportunistic, and it was bold with its machinations. Depending on the era, it tapped certain themes over others, including Marxism as dominant ideology, China's humiliation at the hands of Japanese aggressors, or the importance of a socialist market economy. In 2017, the Party announced that all textbooks would be rewritten to move up the start of the Japanese war by six years. By lengthening the Sino-Japanese conflict to fourteen years and pegging its start to imperial Japan's 1931 invasion of Manchuria, the move was expected to incite patriotism (and even more wariness of the Japanese). Just like that—history was recast.

As China opened up to the world, the Party's target morphed, as if leaders suddenly realized it couldn't realistically cultivate one billion Communists. Instead, they reasoned that a strong national identity would instill a loyalty to the homeland as their people scattered across the globe for jobs and education. In this, there should be no limit: People shall be "influenced" and "nurtured by the patriotic thoughts and spirit all times and everywhere in their daily life," stated the Party's Central Committee in 1994. It was a grand ambition, and the Party's hand reached down not only into the schoolroom but also across film, television, and news media. Patriotism is "thought to be the most effective weapon against the danger of losing Chinese identity and against all kinds of foreign invasions—political, military, economic, and cultural," writes Maosen.

Of course, China isn't the first or even the most fervent nation to try to cultivate a population of patriots. Many American students sing the national anthem each day, and the Fourth of July holiday celebrating US independence from England is a national holiday of parades, barbecues, and flag-bearing patriots. In India, people must stand for the national anthem before watching a movie; many cinemas in Thailand will show a video of their king prior to the main feature and require the audience to stand. Russian high school students must endure military training each year, much like the People's Liberation Army–run equivalent that is required of Chinese schoolchildren of every stripe.

Mused the academic Joel Westheimer:

> If you stepped into a school at a moment of patriotic expression, how could you tell whether you were in a totalitarian nation or a democratic one? Both the totalitarian nation and the democratic one might have students sing a national anthem. You might hear a hip-hip-hooray kind of cheer for our land emanating from the assembly hall of either school. Flags and symbols of national pride might be front and center in each school. And the students of each school might observe a moment of silence for members of their country's armed forces who had been killed in combat.

Here's the distinction: China is singularly unapologetic—and unabashed—in deploying its education system as a governing technique. When the pro-democracy movement Occupy Central brought Hong Kong to a standstill for seventy-nine days, officials in Beijing quickly assigned blame to the lack of patriotic education in the city's school curriculum. "It is clear that there have been problems all along with education in Hong Kong," said China's former deputy Hong Kong official, Chen Zuo'er. Chen then prescribed a cure of "national security and sovereignty" to prevent the growth of "noxious weeds," his term for the students who sat in peaceful protest.

An agenda of persevering patriotism continues in college. China is one of the few countries in the world to deliver a political curriculum inside higher education, with content dictated by the Communist Party.

To me, all this heavy-handedness seemed a bit out of touch with reality, as the Chinese had increasing access to information through the Internet—not to mention counterintuitive. In particular, I wondered how China's leaders could cultivate critical thinking in education—one of many ongoing reform attempts—while also pushing a patriotic agenda? I turned to Beijing academic Xie Xiaoqing. "Might the students begin to question certain elements of their education if they're encouraged to think too freely? Wouldn't the leadership consider this dangerous?" I asked him.

Xie insisted on speaking English, and he became loose-lipped while grandstanding in his second language. "The top leaders hope to develop the students' critical thinking in the fields of physics, mathematics, chemistry, and biology and so on," he admitted, "but *not* in the fields of politics, ethics, and religion."

Not in the fields of politics, ethics, and religion.

That classroom flag defines the parameters for society: Change is fine, as long as those four yellow stars—the Chinese people—are always fixed in orbit around the large yellow star. For the most extraordinary students, official membership in the Communist Party awaits, and the long process of grooming begins early in a schoolchild's career.

So it began with Rainey and his classmates at Soong Qing Ling.

"I get to be *zhirisheng*—student-on-duty—this week!" Rainey told me one day, jumping around our living room.

"What does that mean, Rainey?" I asked, wondering if this would launch his journey down the Red Road.

"I get to put rice in everybody's bowl at lunch," he said. "I add water to the strawberry, bean, and carrot plants. I feed the class caterpillars. I also announce when kids can go outside to play after they finish their morning snack."

Teachers begin rotating children through this student-on-duty position in kindergarten. The purpose is threefold: Teachers get help with classroom tasks, students learn about service to their community, and classmates can take note of how well individual students help their peers. These notes on service are useful when elections for class monitor roll around; in primary, middle, and senior high school years such positions come with power and opportunity, such as the duty of disciplining peers and working closely with school administrators.

I'd always been uncomfortable with the idea that individual students—chosen ones—might have power over their classmates, but in kindergarten such exercises begin innocently. Rainey's teachers ran mock elections for class monitor, and the first time he ran, our son was bested by his classmate, a chipper little girl named Lianpeng. Votes were tallied by body count, and Teacher Song sent by WeChat a picture of the girl's moment of victory: Rainey and Lianpeng standing side by side, with a long line of classmates snaking behind each child.

Finger-tapping the screen, I counted heads: Ten children stood behind Rainey, while thirteen trailed Lianpeng.

Rainey had lost by three. Undaunted, my little boy began immediately strategizing for his next campaign. "Lianpeng had a good speech," Rainey chirped, matter-of-factly. "Lianpeng said stuff that was longer, and more better, and there was more stuff."

A few election cycles later, Rainey finally won by four.

"Mom, you want to hear my speech?" he said, planting in front of our television, words tumbling out tentatively in Mandarin. Every time he forgot a line, he glanced upward, as if expecting the words to rain into his mouth:

> Dear teachers and classmates, I want to be your class monitor. I will serve everyone. If anyone bumps into difficulties, I'll help them solve the problem. If anyone gets hurt, I'll talk to the teacher to ask for help. . . . Please vote for me.

"Great speech, Rainey," I said. "I guess that was enough stuff this time, right?"

"Yes," he beamed. "I said a lot of stuff."

That week, his place in the pecking order was clear. "Because I'm class monitor, I get to tell the student-on-duty what to do," Rainey told me. He brandished a badge he'd attached to his shirt, and high-stepped around our living room to a tune I couldn't hear.

Did President Xi Jinping do the same as a youngster? "It's just play-acting," I tried to reassure myself.

Where I was troubled, Teacher Song was thrilled. "The children work seriously and efficiently, and when they manage order in the class-room as student-on-duty, they become 'little teachers,'" she effused in the Child Development Book. "In this way, the activity can not only correct children's bad behaviors in the early age but also enhance their sense of serving others in the classroom. That's really 'killing two birds with one stone.'"

The cultivation continues in primary school, where all children are encouraged to join Young Pioneers, whose mission statement includes "following the instructions of the Party . . . being successors to the cause of Communism." At fourteen, things start to get serious, and students with the blessing of two League members can apply to the Communist Youth League, akin to a school "for young people to learn about socialism with Chinese characteristics and about Communism, and to serve as a helper and reserve members for the Party," as its constitution states. (Both Premier Li Keqiang and former president Hu Jintao rose through the ranks of the Communist Youth League.)

Today there are nearly ninety million members of the Communist Youth League in China, and another eighty-nine million in the Party proper.

Darcy, my young high school friend in Shanghai, planned to join the Party by his eighteenth birthday.

At seventeen, Darcy was already one of the chosen ones. "I am a *jijifenzi*—and my plan is working," he told me, using a term that means "zealot" or "enthusiast," and is also a status designated by the Party. By his junior year in high school, teachers in his grade had vouched for him, and he was chosen to start the process: He'd already been given initiation rites, attended special classes, and written reports praising the Party. It was an invitation bestowed upon only three of four hundred students in his grade level.

He showed me a letter he'd penned as part of his application. The title was "Walking on the Red Road":

They are brave, regardless of war or disaster.
They always run ahead, throwing your head to shed blood.
They are wise, with farsighted leadership of a new China.
They are walking on the red road people, they are Communists!

Darcy's list of accomplishments revealed a distinguished Young Pioneer with the right kind of grooming to serve in the league—disciplinary board member, discipline inspector, vice squad leader—and the boy concluded that he would work to secure the long future of the Communist Party: "I set foot on the red road, in front of the older generation of revolutionaries, and we will try to catch up, take the stick in their hands, overcome all obstacles, so that the red road is wider and longer!"

I gazed at the baby-faced teen who sipped coffee before me. He wore an Adidas tracksuit and Nike cross trainers, apparently the costume of today's budding Communist leader. "Do you believe everything you wrote here?" I asked, fingering the paper.

Darcy paused. "Yes, I believe," he said, and nodded, wall of bangs moving with his head.

As we became better acquainted, Darcy's House of Communist crumbled under scrutiny. Much of what I've written is official *guanfang*—gobbledygook, Darcy told me once, in a whisper. "Just stuff I have to do," he said. "It's what you need to do to get ahead." His ultimate goal was outside the reach of the Party and off the trail of the red road: graduate school at the University of Michigan.

"You want to study abroad?" I asked him.

"Yes," he said, and nodded, as a worrisome thought suddenly entered his brain.

"If I want to study in America, will they care if I'm a Communist?" he asked me, head bowed but eyes fixed on mine, waiting for an answer.

I realized that the face my friend presented to the world flipped and flopped depending on the audience and the purpose at hand. But one thing was clear: As an ambitious child of migrants, Darcy would join the Party because it ensured a brighter future inside Chinese borders. Membership opened up a new world of perks: eligibility for scholarships, the networks to find good jobs, positions at state-owned enterprises. Party membership is required for promotion in many government and university jobs.

Darcy was among China's best and brightest, and these were things he simply had to do. As I pondered the futures of my two Chinese student friends, I realized they were set for sharply divergent paths.

Whereas Amanda's year abroad would pave the way for college overseas, Darcy must navigate Chinese society the best he could.

⋆ ⋆ ⋆

WHEN I THOUGHT about Darcy's situation, I realized most Chinese present public faces that are very different from their private thoughts. To me this state of affairs seemed particularly problematic for China's youngest generation.

As an American, the self I offered to the outside world and the one I paraded around in the privacy of my apartment were mostly one and the same. Sure, a fully functioning member of any society must follow its spoken and unspoken rules, but for the most part, I could offer any thought I wanted for all to see.

Darcy didn't have that luxury. When I looked a little bit deeper into this issue, I discovered I wasn't the only one who worried: A handful of brave Chinese scholars criticized China's moral education curriculum for this very reason, assigning the Party's heavy hand a losing proposition and a failing letter grade.

Here's the problem: A huge gap exists between what kids are taught in school (government provides for you) and what they see happening in real life (government officials taking privileges for themselves). There's an immense gulf between what people *say* they believe ("Long Live the Communists") and what desires they harbor in their personal lives ("Capitalism affords me opportunity, and a decent pair of Nikes").

China's speeding economic change has shaken the socialist values upon which the Party was founded. Capitalism abounds, and spending hours in the classroom trying to convince students otherwise has become "unconvincing and unreliable in many respects," writes the moral education scholar Li Maosen. The Party's attempts at moral education have simply become a "means of political indoctrination for the purpose of ruling the people rather than for the development of the individual person." The public intellectual Ran Yunfei derides a society "where the educational materials are all about loving the Party—of course it leads to a spiritual crisis," he told the *New York Review of Books*.

China is growing a nation of patriots who worship the Party in public but cultivate alternative thoughts in private.

Take the models that the Communists put forth as heroes. Darcy and Amanda, having gone through nearly two decades of a public Chinese education, could rattle off their names and stories without pause: Lai Ning was a fourteen-year-old student who perished while assisting firefighters in tackling a blaze in Sichuan province. He was remembered as a standoffish yet studious schoolboy, but after his death, the state declared him a selfless, national hero. Huang Jiguang used his own body to block enemy bullets while struggling bravely for the Chinese during the Korean War. Soldier Dong Cunrui fought for the Communists in 1948, and he detonated explosives to clear a bunker. "For a new China!" he yelled, as the story goes, just before sacrificing himself with the act.

Disseminated through the curriculum and presented in class, these models are almost always People's Liberation Army soldiers or devotees who exhibited extraordinary devotion to Mao Zedong, Party elders, or the greater good.

Most Chinese today believe their stories are out of touch with reality. And it's no surprise that any tale encouraging a generation of precious, only children to self-sacrifice might not sit well with parents.

In private, families have their own interpretations. A girlfriend put her two girls in public school in Guangdong. One week the young students were told the story of a man who carried a Chinese flag with pride but who was also slowly starving to death in the hot sun. One day he encountered a potential savior: a passerby with a loaf of bread.

"I'll trade you this loaf of bread for that flag," the passerby said.

"No," replied the starving flag-bearer, despite the hunger in his stomach and the weakness in his limbs.

"How about ten loaves for the flag?"

"Still, no," said the flag-bearer.

As the story goes, the man died of starvation, holding the flag upright as long as he could bear it. The teacher would then echo the moral of the story: "How brave the man was!"

My friend was horrified. "I try to combat the brainwashing at home," she told me. "I teach my kids to take the bread! Right, girls?" she said, looking over at them during brunch at our home. "What do we choose instead of the flag?"

"The bread!" both girls, ages six and nine, answered in unison.

* * *

THE TENSION BETWEEN ideology and reality was on full display when I visited a high school political science class, curious to understand whether Chinese students recognized the Party's efforts to indoctrinate them.

"You are the future owners of this country," Teacher Qiu told her thirty-two Shanghai high school students. "You are the future of the motherland, and our hope. If you travel down the wrong path, you all lead our country to nowhere."

I'd slipped into the classroom and sat in the back, invited to observe by the principal, who was acquainted with my research assistant. The students were parked in pairs at metal desks, which were in turn grouped into three columns, running from the chalkboard to the back of the room. I'd always been struck by the inside of the Chinese classroom: daylight glaring through single-paned windows, lone circular fan descending from a high plaster ceiling, green chalkboard cleaned by students during break, the etchings of the previous day's instruction still faintly detectable. It felt more like a barracks than a classroom.

Mounted on the front wall was the Chinese flag, and the students sat attentively.

"Now, let's talk about the civil service exam," Teacher Qiu said. "Please speak your thoughts. Why do thousands of people try to get onto that boat of civil service jobs?"

"Because civil service jobs are an iron bowl—a secured job."

"A *gold* bowl," another student exclaimed. "A super-secured job."

"Money!" chimed in one student, a shudder passing through her shoulders. It was wintertime. The students wore heavy red-and-gray coats over their uniforms—public schools don't usually have heat—and shivered over their textbooks.

"Not much money but good stability! Good welfare!" another echoed.

Another boy chimed in, his voice low and husky. "They can also embezzle."

"Embezzlement?" Teacher Qiu repeated, echoing my surprise. In recent years, foreign investigative journalists had uncovered billions of assets sitting in accounts and shell companies held by family members of sitting Communist leaders. It was money derived from business contacts, resulting directly from a connection to power. This was an open secret, but I hadn't expected students to feel comfortable openly questioning Party leaders in a classroom.

Teacher Qiu looked at the boy, who rephrased. "To put it more mildly, to earn extra money," he said.

"Don't babble," Qiu admonished, nodding toward me, where I sat at the back of the room. "You are embarrassing the teacher who is sitting in the class." It was a gentle reminder that outsiders sat in the room: Don't pull back the curtain too far.

The teacher tried to redirect. "Public servants are just like us. *Anyone* can take the exam and become a public servant, fulfill their obligations, and enjoy the security provided by the state. They are not to be envied."

To be quite honest, I wasn't sure who was more skeptical of her statement: the teacher or the students. Chinese society has become a brutal, dog-eat-dog environment where everyone is obsessed with getting ahead; in such a competitive environment, a lecture about serving the state rings hollow.

A different student spoke. "Public servants," he announced, "become the *privileged* class, which contradicts their role as people's servants."

The boy was right to criticize. A financial and ethical morass—assisted by a lack of transparency or due process in government—had cast a shadow on China's government dealings at nearly every level. The Western equivalent of the Chinese central government corruption recently uncovered might be a skyscraper sitting in the name of the UK prime minister's cousin, or the discovery that a US president's mother owned stakes in insurance companies worth $800 million—despite the fact that neither the cousin nor the mother worked in real estate or insurance. When this happens in China, the people have no recourse.

The teacher challenged the student. "*Most* public servants have no privileges. Most adhere to proper procedure—in proper order. For example, when we get up in the morning, we must first wash our faces and then brush our teeth, and then can we go eat breakfast."

"But some eat breakfast first," another student chimed in. I imagined this was his euphemism for officials who embezzle public funds or leverage power inappropriately.

"Sure, some eat breakfast first," Teacher Qiu said, trying to steer the conversation back on track. "*Then* they wash their faces and brush their teeth."

"Your example is not appropriate," the student countered.

"Do you have other examples?" the teacher asked, glancing my way. "What kind of behaviors require a certain order? For instance, handing in homework?"

"Assembly lines in a capitalist society," the student responded. This was also a challenge. Capitalism is a no-no; the official Party line for China's system is "socialism with Chinese characteristics."

"What kind of assembly line—in what kind of order?" Teacher Qiu posed, working again to redirect.

The student responded with a chant. "Long live Communism, and long live the Communist Party." He chortled, and the whole class tittered.

"You shouted out a slogan," Teacher Qiu said. The student offered another one, the sarcasm drowning his voice. This time, he used the slang for President Xi Jinping.

"Xi Da Da is a good person."

* * *

IN PRIVATE, Teacher Qiu spoke of a simpler time, hands folded in her lap.

Public servants used to inspire the admiration of students, Qiu told me, class dismissed, the nostalgia of another era softening her voice. "This admiration was unequivocal. Today, students challenge the notion that 'domestic shame should be kept hidden.'"

And China's leadership prefers 'domestic shame' buried under ten feet of Yunnanese soil, preferably frozen solid to deter ice picks and shovels. Civil servants such as Teacher Qiu are expected to be infantrymen in the effort to keep the populace in line.

Qiu longed for another era, but a lanky, charismatic history teacher I met in Beijing was more than content to sweep burgeoning sentiments of dissent under a rug. Teacher Kang never questioned the boundaries of what can be discussed in the classroom.

"Every Chinese knows what can be said in class and what cannot," Kang boomed. I'd met him at an education conference in Beijing, and he agreed to talk with me about moral education. We'd bantered for about an hour about theories of civil education before he began to open up.

"Is there an instruction manual to what can't be said?" I asked Teacher Kang, half joking.

"There's nothing clearly written," Kang responded, although a 2013 central government document was explicit: It banned discussion of democracy, freedom of speech, and past mistakes of the Communist Party.

"Then how do you know?" I asked.

"I just know it," Kang continued, matter-of-factly. "Nothing is *mingwen*—explicitly stated."

I pressed. "What, for example? What can't you say in class?"

Kang rattled off a list. "You should not mention Falun Gong. You should especially not mention Falun Gong. That's more sensitive than the June Fourth incident. Also the Islam problem." Falun Gong is a Buddhist spiritual practice deemed a threat to social stability, and "the June Fourth incident" is shorthand for the June 4, 1989, Tiananmen Square massacre, in which Beijing sent tens of thousands of troops to crack down on a student pro-democracy demonstration. The "Islam problem" refers to Muslim separatists in Xinjiang and their perceived threat to the Chinese Han majority ethnic group.

"It is possible to criticize the government?" I asked Kang.

Kang thought for a little bit. "I can't give you a simple yes or no answer, it depends. You can discuss a specific problem of government, such as Beijing's terrible urban planning, our traffic jams, and awful public service."

"What about corruption?"

"Corruption can be discussed. This is no problem," Kang affirmed.

"Are you sure?" I said. This was surprising to me, but it explained what I saw in that political class.

Kang clarified. "You cannot challenge the legitimacy of the Chinese Communist Party. This you should avoid. But you can discuss the *causation* of corruption—for example, the lack of checks and balances of power."

"How do you know where the line is drawn?"

At this, he laughed heartily, his head jolting to the right. "I am Chinese. I know. History tells you this. The Cultural Revolution was a dark time and China learned its lesson." This was Mao Zedong's ill-fated attempt to purge capitalist and traditional thought from Chinese society and to reinstall Maoism as a central philosophy, a campaign that resulted in the persecution of millions.

A Harvard-based researcher backed up Kang's assertions about speech; Gary King analyzed nearly eleven million social media posts and found that it's a call to action—rather than language—that Beijing is most interested in suppressing. "Words alone are permitted, no matter how critical and vitriolic," the researchers wrote. "But mere mentions of collective action—of any large gathering not sponsored by the state, whether peaceful or in protest—are censored immediately."

Recently, President Xi Jinping has tightened his grip on academic freedom—an action one scholar likened to a "minor cultural revolution"—and also tapered the channels that bring the content of Western curricula into China at all levels of schooling.

I asked Teacher Kang about the Party's increasingly heavy hand.

"Setbacks are inevitable," he admitted, but he also said that an authoritarian government is necessary for China in this stage of development. "A civil society hasn't yet formed. It's not rational to build a democracy for China, simply because a democracy is more advanced."

My Chinese student friend Darcy, in fact, believed a two-party system was patently inefficient. "With one party, policies can be implemented," he said, with a firm nod. "In America, if the Democrats wanted to ban guns, the whole nation would ban guns. But because Republicans object, it's hard to enforce. In China, there's only one party. They can do what they want immediately." During other conversations, Darcy had hinted at differences between his public and private faces, but on the benefits of a one-party system, he seemed patently certain.

Amanda had a similarly practical approach. As a lead student organizer for a Model United Nations conference in China, she'd been told certain topics could not be discussed: "Taiwan and Tibet, the election in Hong Kong, race in South Africa, and the conflict between Iran and Israel. Our director tells us what we can't talk about."

I was flabbergasted. I expected outrage, or at least a little bit of displeasure at the fact that Amanda was told what she couldn't discuss. "So—the purpose of Model UN is to encourage freethinking among the next generation of young leaders," I challenged her. "You're telling me about outright censorship of debate topics, inside an event whose purpose is to . . . debate world issues!"

Amanda shrugged me off. "We cannot afford to care about censorship—that's a Western luxury. You forget that China is still a developing country and we are still focusing on making sure we have food, housing, and education."

As she saw the incredulity growing in my eyes, Amanda put me in check with an observation gleaned from her year in the United States. "Americans have this illusion of freedom and democracy," she told me, "but it was the Founding Fathers who created the Constitution. An elite group controlled by a minority."

Good point.

"America is not truly a democracy—it's elitism," she declared. "By telling people it's a democracy, it gives people an illusion that there's hope. That's the difference between Chinese and Americans—the Chinese are devoid of real hope, so they just concentrate on their personal issues."

With that, she looked down into her coffee, resolutely.

Our discussion was finished.

★ ★ ★

ONE WEEK I caught Rainey singing "I'm a little soldier, I practice every day," in Mandarin, while marching down the hallway. His arms swung to and fro as his knees lifted to the ceiling.

I take a wooden gun—bang, bang, bang!
I drive a small gunboat—boom boom boom!

At the word "gun," Rainey clicked left hand to right elbow and crouched forward. His right hand made the shape of a gun, as if he were an infantryman on the lookout for enemy fire.

I ride as a cavalryman—go go go!
I am a little soldier, I practice every day.
One-two-one, one-two-one, Let us forward march!
FOR . . . WARD . . . MARCH!

"That was wonderful, Rainey," I called after his retreating backside. I blinked, hard, as he marched into his room.

Earlier in the year, he'd also trumpeted the lyrics of "The East Is Red": "The East is Red, the sun is rising. From China comes Mao Zedong. He strives for the people's happiness. Hurrah, he is the people's great savior!" At the time, I polled American and European friends whose kids were also in local Chinese school, and they seemed to fall into two camps: They were either repulsed or they shrugged it off. "What do you expect? It's a Chinese school!" my German friend Chris had teased me, before zooming off on his scooter, heading to work.

The following week, as Rainey and I headed out the door for school, I glanced down to see my son clutching a thin black trash bag that he'd salvaged from underneath the kitchen sink.

"What's that bag for?" I asked him.

"It's for trash," he told me.

He skipped ahead of me on the walk to Soong Qing Ling, and I watched as he stooped to collect detritus inside the grounds of our apartment complex. He rooted around in the leaves to find a snapped twig, a pebble, and a discarded lollipop wrapper. He inserted each into the plastic bag gripped in his left hand.

On the street, the trash got sinister: He spotted a broken glass medicine bottle, the type that awaits a syringe. It appeared to be a used vial of diabetes medication.

"Uh . . . I'll take that one," I said, swooping it out of reach. I inserted it in the bag.

In that moment he spotted a crumpled tissue, which he coaxed into his bag.

"Rainey, can we please stop? What are you doing?"

"I am making China beautiful," he responded.

"Why are you making China beautiful, Rainey? Who told you about that?"

"The teachers."

At various moments on that walk to school, Rainey bent and stooped, spotting more treasures for his trash bag. As we approached Soong Qing Ling's iron gates, a grandmother saw Rainey's bag, and, as many Chinese often do, she spotted a teaching moment for her progeny. She rapped her grandson on the back of the head. *FWWHaaap*!

"Look at him, he's collecting trash. He's a *haohaizi*—a good kid," she said. "You're a bad kid, you don't pick up trash." She delivered a firm push on his shoulder, and the child lurched forward.

Before we reached the classroom, Rainey handed me the bag. I peered in and saw a montage of China: The old and the new, the used and the discarded. A peanut. Dirty leaves. A dried lotus flower. An ice cream wrapper. The gold foil of a Ferrero Rocher chocolate. The glass medicine bottle with the characters for "insulin."

We arrived at the classroom, and I told Teacher Song about Rainey's

China beautification campaign. She beamed and patted him on the head. "Rainey, why don't you put this in the trash can now."

Later, I dialed Rob. "Our oldest son is slowly morphing into a worker for the state," I panted into the phone, as a group of slogans Amanda learned during her school days echoed in my head:

劳动最光荣: Labor is the most glorious thing.
无私奉献, 舍己为人: Sacrifice unselfishly for others.
报效祖国服务社会: Serve your country, serve society.
人民的利益高于一切: The interest of the people is higher than all else.

"That's ridiculous," Rob told me. "It's good that Rainey picks up trash."

"Yes, of course it's good," I told Rob, "but I feel like we're losing control of his mind . . . the school's become too powerful." Whether my son might eventually want to join the Communist Youth League wasn't a concern of mine. Until now.

Rob laughed. "Don't worry about it," he said. "He has us at home."

I broached the topic with Darcy, who also shrugged off my worries. "China would never truly be a united front of patriots," Darcy said, voice low, in an unguarded moment. "Nationalism is all more in theory," he said. "In real life, human nature makes people consider themselves."

Amanda's conviction was even firmer. "The curriculum is all bullshit! Brainwashing! On the surface teachers praise socialism. The Party members need to write reports. But everyone knows this is not a socialist economy. It's obvious! It's obvious to everyone!"

School may have a student's attention for the moment, but try as it may, it's clear that the powers that be cannot control context, critical thought, or anything that proliferates inside the privacy of one's own head. China may conceive of a nation of patriots, and inside a child's educational journey there may be times that toeing the Party line may seem expedient—or even necessary. But the Chinese I knew recognized the Party's blatant attempts to censor and control access to information, and, ultimately, Amanda, Darcy, and the moral education scholars were convinced that we had nothing to worry about.

China's grip on the hearts and minds of its people would only grow more tenuous.

* * *

THE FOLLOWING WEEK, Rainey was drawing over by the window, working with blue and gold markers. I peeked over his shoulder. He was sketching what looked like a majestic, tiered monument. Several dozen police figures dotted his square, drawn with ovals for heads, atop squatty bodies, with thin scrawny lines for arms and legs. A squarish military hat sat atop each oval.

"What's this, Rainey?" I said, sweeping my hand over the drawing.

"You don't know Tiananmen?" he said to me, his voice full of surprise. I looked at the picture. In the middle of the building, he'd placed the portrait of a man with no hair—an oval for the face, two dotted eyes, and a mouth upturned at the edges.

"Who's that?" I asked, pointing to the man, though I already knew the answer.

"That's Mao!" Rainey said. When most Westerners hear the word "Tiananmen," they think of 1989, the lone tank man, a student democracy movement, and the brutal government-sanctioned massacre of protestors in the square.

Not my four-year-old boy. To Rainey, Mao Zedong was a smiling egghead and the square a monument to Chinese excellence.

"What did the teachers tell you about Tiananmen, Rainey?"

"It's a place in Beijing," he said.

"What else?" I asked.

But Rainey had exhausted his memory on the topic.

Of course, the teachers had said nothing of its importance in world history as the site of a democratic movement or a subsequent, brutal massacre. It was an event that drew the world's scrutiny on Chinese human rights, and the student demonstration might have changed history by instigating a rebellion against the party in power. But the teacher wouldn't have said that. No teacher in China dared to discuss it, and, in fact, they couldn't even refer to the event by its proper month, day, and year.

Instead, Rainey's teacher might have mentioned the square as a monument to Mao Zedong, or its representation of civic duties: country, people, labor, science, and socialism.

As I glanced down at my son's drawing, I instinctively felt that whatever Beijing's desire to reform education, its basic purpose wouldn't alter course for years, if not decades. The preservation—the legitimacy—of the Communist Party depended on it.

I peered closer to examine my son's paper policemen and noticed a stark, dark shape in the place of some officers' hands.

Rainey had drawn half of the stick figures with guns in hand.

8

ONE HUNDRED DAYS 'TIL TEST TIME

I have traveled across more than twenty countries all over the world, and not a single one of them has as big a gap as China. The gap is one hundred years in economics, ideology, and concept.

—ZHOU NIAN LI, EDUCATION PROFESSOR AT EAST CHINA NORMAL UNIVERSITY

When he isn't playing soccer or studying Chinese characters, Rainey likes to draw.

At a small, blue children's table underneath our dining-room window, my little boy's touchstones materialize on paper: birthday presents trussed up in bows, pork dumplings, sunfish caught in Minnesota lakes, Mao Zedong and the Tiananmen Square police.

One afternoon, Rainey traced out a thick, single black line that took the shape of a flightless bird.

"What's that, Rainey?" I asked.

"That's China," he said, assuredly, continuing to move his marker.

"That's not a rooster?" I pressed.

"No. That's China," he said, continuing to draw.

Rainey's China very much resembled a rooster. I watched carefully as my son's imagination emerged on sketch paper. This rooster is full-bodied with a chest puffed in conceit, a head and cockscomb that jut north into Russia, a curved back that cradles Mongolia, and imaginary feet that plant somewhere in the South China Sea. The capital city of Beijing sits on the eastern seaboard right at the throat of the cock, in prime position to exert a chokehold over the rest of body. I tapped on this center of power absentmindedly.

"Where is Shanghai, Mom?" Rainey asked.

"Right here. We live right here," I said, as I traced my finger south along the coastline to find Shanghai situated at the crest of the rooster's proud, puffed-out chest. It was a fitting spot for China's trophy, skyscraper city.

"Where is Ayi from?" Rainey asked, recalling that our nanny hopped the train during holidays to a faraway place to see family.

"Jiangsu province," I said, pointing my finger north of Shanghai, in the rooster's neck.

Almost in tandem, Rainey and I turned our gaze west over the body of the rooster, taking in the expanse of land that ran toward the borders of India, Pakistan, Kazakhstan, Kyrgyzstan, Tajikistan, and Nepal.

"What's this here?" Rainey asked, placing his palm flat on the page.

"That's western China, Rainey," I said, waving my entire hand over the body and tail of his rooster. Most foreign visitors travel only to the major cities concentrated in the east. As Rainey's finger crept toward Tibet in the middle of the Asian continent, I thought about the hundreds of millions of Chinese who live west of China's wealthy, urbanized coast, in a mixture of large cities, smaller counties, and rural areas.

What were their schools like? Did rural students feel academic pressure as the city folk did? "Spend time outside Beijing and Shanghai," urged a fellow Stanford graduate who'd taught for two years in rural Guangdong province. "Visit rural China; the difference is night and day," other educators told me. Another put it succinctly: "Understand where China's school system came from, and you'll better understand where it's headed."

Conditions in rural schools could be dismal. It wasn't unheard of to find 130 kids jammed into a single classroom, a lack of potable water, and ill-trained teachers. Children were placed in boarding homes or schools while their parents toiled in faraway cities, with often-illiterate grandparents left to help with homework. Others might be fortunate to share the same roof as their parents, but the average child's caretakers knew nothing of raising a child to excel in modern China. "They knew more about raising a pig to slaughter than nourishing a child," read the report of one American research group.

When Shanghai's chart-topping students were announced to the world a few years ago, whispers about rural China intensified. Those world No. 1 PISA scores don't include rural areas, said critics who questioned the usefulness of the test. Tally in students from poorer, rural regions—the body and tail of the rooster—and China's average score would plummet like a bag of stones tossed into the Yangtze River, they insisted.

One week, I stumbled upon some staggering numbers.

Nearly half of all children *outside* of China's large cities are high school dropouts. In rural areas, only about seven of every hundred Chinese attend university of any kind; children in cities are almost twelve times as likely to go to university. Income studies show a wealth gap between rich and poor China as large as the chasm between London and Bangladesh. As China's economy rocketed to the moon over the last few decades, it had left behind entire colonies of stragglers. How would the country continue to modernize and grow if hundreds of millions of poorly educated, impoverished Chinese languished in the countryside?

I glanced down at Rainey's drawing.

On his sketch paper, the distance between rural and urban China was only the width of a fingertip or two.

In reality, that distance would make all the difference.

★ ★ ★

I MET LAUREN in Shanghai in 2012.

Lauren was a migrant from rural Anhui province just west of Shanghai, commonly called one of the poorest provinces in China. If we happened to see each other shortly after she'd gone home for a visit, she always brought me a shopper's plastic bag full of eggs—freshly laid orbs checkered with hen poop—which she urged me to steam and feed to my family.

"It's good for eyesight," she'd tell me, head bobbing with enthusiasm. "Our countryside eggs are nutritious—see the bright orange in the yolk? I sat with this bag in my lap on the bus for eight hours, and not a single egg broke!"

Lauren had wide shoulders and hips, broad feet made for patting down earth, and short thighbones that helped maintain balance during squatting. Hers was a peasant's body, built for working the fields, but she'd left the countryside twenty years ago to chase big-city money. After cycling through jobs in factories and restaurants, it finally dawned on her: The heft of a body built for the countryside would be great at delivering massages in the city. So she trained as a masseuse. For every hour she spent in dark bedrooms in Shanghai she put 75 RMB—about $13—in her pocket. She moved silently, kneading backsides with agility and speed, and her hands rarely tired during the job.

"My real name is Long Jiang," she told me when I first met her, "but a German client said 'Lauren' would make me more approachable to foreigners."

Lauren was a practical woman, and if a perfect stranger thought an Anglophone name would help her make money, she would take his suggestion. She'd always had that countryside practicality, and when she'd birthed her only child fifteen years ago, she chose a utilitarian name for him, too.

"*Jun*军, or Soldier," she told me. "Being a soldier is a good job. It's the iron rice bowl—a stable government job. You get a pension."

There's an unspoken Chinese rule about baby naming: It's perfectly fine, even encouraged, to choose a name that mirrors exactly what your heart desires. There's no need to disguise ambition, hope, or even greed. I've met many Chinese who have named their children after heady occupations, ambitions, government leaders, or even physical characteristics that society deems desirable: "Big Boss." "Tall." "Rich." "Beautiful." The latter, "*mei*," is a very popular given name for Chinese girls. Celebrities, pop stars, and entrepreneurs aren't off-limits, nor are foreign luxury cars.

I volunteer-taught English classes for two years at a Shanghai kindergarten, and one of my favorite students was nicknamed "Ferrari."

"Why did your parents choose 'Ferrari'?" I'd asked the five-year-old boy one day.

"Because my mom said I'd own one someday," the boy responded.

Anglo naming rituals employ more restraint, with most given names having some kind of biblical or historical root; in other words, you must crack open the King James Bible or ponder Latin origins to get at a deeper meaning. Our sons' Anglo names mean "strong counselor" and "broad hill," but you'd never know it unless you went researching English and Germanic root forms.

When it came time to give our sons Chinese names, I knew I'd have to quash my desire for subtlety. A long-held Chinese tradition grants naming responsibilities to the male elders of the family, and of all the trends I'd bucked, this one would be easy to follow. I asked my father to choose.

"I will start research immediately," he told me. After consulting ancient texts, a Chinese dictionary, and a writer friend, my father announced, "Rainey's Chinese name will be 磊." Pronounced "*lei*," it was the character for stone 石 written three times. Stone times three. It meant "open" and "honest," and I was thrilled that he'd chosen a name with positive character traits that were unrelated to money or prosperity.

For Rainey's little brother Landon, born three years later, my father got ambitious. "鑫—the character for gold 金written three times," he proclaimed. "It is pronounced *xin*." Gold times three.

"It means 'profit and prosperity,'" my father wrote me from his home in Houston.

"We feel weird naming our kid after wealth and fortune and prosperity," I wrote to my father from Shanghai, belly bulging as I entered my last weeks of pregnancy. "Are you sure it's not obnoxious?"

"The character for 'gold' alone is kind of 'obnoxious,' but with three golds together, it's not. That's the amazing thing about the Chinese language," he wrote back, diplomatically.

"Fine. 'Profit and prosperity' it is," I said. Gold times three. Perhaps our son's given Chinese name would steer him into a higher-earning profession than journalism or writing.

For Lauren, her son's name served up a daily reminder of the goal she'd set for him: Joining the People's Liberation Army would spare the boy a lifetime of manual labor. Lauren had another hope: She wanted Jun Jun to be able to sleep in the same bed with his future wife. Because Lauren and Wang were migrant laborers, their jobs took them to different cities, linked only by mobile phone and long-distance bus or train. Lauren's singular mandate was that her son avoid the fate of his migrant worker parents.

Neither had much of a formal education. Lauren's parents had sent her to the village schoolhouse three years late, and her educational decline was swift: The eight-year-old was simply too far behind to catch up. Before long, she found herself sitting behind the five-year-olds in the room, the teacher's words gibberish to her ears. When the new term started the following fall, she simply stayed home. In his town about a mile away, Wang fared three years better, finishing the fourth grade, but he dropped out after his family failed to come up with fees for meals, textbooks, and other incidentals.

Lauren and her husband had met as teenagers working in Hangzhou, when they were both "out" from their villages. Wang was wiry and quick on his feet, and Lauren immediately sensed he'd put his restless energies toward work. When Lauren first brought her prospect home, her father proclaimed loudly, "He has no money, no house, and no car," as Wang eyed the floor. True, yes, but Lauren refused to consider other men. After the wedding, the pair went back out again, she to a clothing factory and he to a construction site where he ate, worked, and slept. They met during holidays, and soon enough, Lauren gave birth to little Jun Jun. After a few months, Lauren left the boy with her parents and headed back out, milk leaking through her shirt for days, she recalled.

She'd made the right choice in Wang, Lauren told me. Wang doesn't pay for sex, a temptation that befalls many migrant men who see their wives only once a year. There was plenty else to celebrate. After fifteen years working in separate cities, they'd saved enough money to build a four-story home in the countryside, resplendent with tile finishings and a working refrigerator.

"*Bu cuo*—not bad," Lauren always told me. Economically and romantically, they were a success story in China's rural-to-urban migration, which increased the population of Chinese cities by roughly four hundred million over the last three decades.

When it came to Jun Jun's educational story, Lauren's fortunes shifted. In Shanghai, Lauren's massage business was thriving, but back in Anhui province, Jun Jun was struggling. China's antiquated household registration system—*hukou*—kept Jun Jun from enrolling in public schools outside his home district. That meant that he couldn't join his mother to attend school wherever she migrated for work. For the first fifteen years of his life, the boy lived with his grandparents while Lauren was out working, but for middle school, she moved him to a boarding home in nearby Jingxian County. It was a town of 350,000 with good schools and higher graduation rates.

The teenager found himself suddenly surrounded by strangers. Conditions at the boarding home were something out of a Charles Dickens novel. The cook washed vegetables in a basin with her feet. Jun Jun slept eight to a room, in bunk beds stacked two and three high to the ceiling, sharing one apartment with about forty children. Lauren was paying the boarding mistress a fortune by Chinese standards—4,900 RMB ($800) a semester—but Jun Jun was constantly hungry and didn't have enough water to drink. The glass had been punched out of the windows, leaving only vertical black bars. There was little light by which to read or study, and other teens kept him awake by talking late into the night.

Exhausted without a good night's sleep, Jun Jun had begun falling asleep during class.

Lauren could pay for his educational needs—food, books, tuition for private or experimental schools, outside tutoring, teacher gifts—but what Jun Jun really needed was a hug from a family member and real adult supervision.

"He's become a *liushouertong*, a 'left-behind child,'" Lauren moaned. At the time, a fifth of China's children—sixty million kids under the age of eighteen—lived with one or no parents, casualties of China's push to move working-age rural Chinese into big cities. Like any child without the discipline a watchful parent provides, Jun Jun had become addicted to video games. He cared nothing of study, and he was depressed.

The National High School Entrance Exam was just a year away. This *zhongkao* would determine whether Jun Jun would go on to a traditional high school, with the eventual hope of going on to a proper college. For most Chinese, this exam is of far greater importance than the college test, and it determined whether a family might join the ranks of the emergent middle class.

"Jun Jun must get into high school. He must go to college," Lauren kept repeating, the image of Jun Jun as a migrant worker her worst nightmare.

One day, Lauren got into a fight via text messages with the headmistress. Jun Jun had suffered a particularly bad showing on a test.

"My son's grades have fallen by half since I placed him in your home," Lauren had messaged to her son's boarding mistress Zhang, who had promised free tutoring that never materialized. "I hope you will take responsibility for him just like you do the other children."

"How can you send me a message like that?" Headmistress Zhang spat back. "You have no education, and you have no *suzhi*—quality."

This was the ultimate insult to an uneducated migrant. "You are right, I have no education. But I *do* have quality," Lauren spat back, tapping out a message as quickly as her thumbs could move.

"A mother like you, no wonder your son is the way he is. All the other parents take me out to eat. They express their appreciation. You don't," Zhang wrote.

"I'm paying you to take care of my child. How can you talk to me like this?" Lauren retorted.

But Zhang didn't respond, finding a more certain target the next day: Jun Jun. The headmistress slapped the boy across the face, jeering, "Your mother doesn't love you."

Lauren was in Shanghai when her son relayed what had happened, and she became frantic. Jun Jun was in this woman's care and Lauren was six to nine hours away by long-distance bus.

"Jun Jun must be brokenhearted," she told me, fretting over what to do next. "He might start to believe we really don't love him. After all, we left him there, didn't we?" she said.

Lauren considered filing a grievance, but Zhang's operation was illegal and no authority would act on her complaint. Jun Jun was approaching his last year of compulsory education, and the National High School Entrance Exam sat sentry at the end of that road. His grades were slipping, along with the dreams Lauren and her husband had long held for him.

It was time to go home.

* * *

LAUREN INVITED ME to visit her in the countryside that spring.

Since she'd left Shanghai a year earlier, I'd received a trickle of text messages that indicated a family trying to build a new life together, after a decade and a half apart: "We rented an apartment. The floor is unfinished, but it's close to school."

"Jun Jun hates study. I don't know how to help him."

China had developed the high-speed rail and built the world's longest subway system in just over a decade, but the best way to get to Jingxian County was still a long-distance bus, one of those crawling, diesel-powered carriages that were a throwback to simpler times.

I stepped up three giant, knee-high steps. My fellow passengers were dozens of migrants headed back home, and they munched on nuts and crackers, making popping noises with their teeth, tossing refuse in blue plastic bins that dotted the aisle. They coughed up phlegm with hacking, guttural noises and spat it in the aisle. My game of bladder-hold began; there wouldn't be a pit stop for two hours, and even then, relieving yourself required squatting over long troughs of running water at a rural rest stop alongside other travelers, pants around ankles.

"The roads are good," the driver announced in Mandarin tinged with a rural accent, "so today should be only a seven-hour journey."

As we exited the sleek, elevated expressways of China's trophy megacity, the skies grayed and the buildings shrank. Ten-lane highways became two-lane roads that sagged with dirt and potholes, with patches that were hastily poured. Our bus passed crews working alongside the road, clearing lots for shiny new buildings. The lots started right at the edge of the highway, and past that there was nothing as far as I could see.

The tableau that unfolded outside the window was one of unblinking, rapid transformation as fallow land disappeared under the crank of yellow-orange excavators. From my seat on the bus, I was witnessing the ambition of countryside that wanted desperately to be city. Billboards erected along the way hovered over giant mounds of unearthed russet-colored dirt, advertising progress through slogans:

Buy a solar heating machine.

A trusty bank is a money tree for generations.

Build a civilized society and promote leaping developments.

Building "leaping developments" was a dirty affair, and for much of the journey to Jingxian, construction dust hung in the air, clouding the sky and settling on the trees, shrouding their green leaves in a fine brown layer of silt. Looking out the window made me thirsty.

The National High School Entrance Exam was just three months away. A full year had passed since Lauren's text-message fight with Jun Jun's boarding headmistress. After fifteen years living apart in three separate cities, the entire family was back under the same roof: father, mother, son.

I fell asleep, my head jammed against a tall picture window, thoughts bouncing around as jarringly as the bus on the pitted road. Jun Jun grew up with little supervision, and suddenly his parents slept in the bed next to him. How was the teenager adjusting? What were his prospects on the *zhongkao*? Lauren and her husband were uneducated; how could they expect to help Jun Jun in math and physics?

Hours later, my bus rolled into Jingxian station, and I spotted Lauren through the dust-covered windows. She was standing before a navy blue Baojun sedan, a Chinese car. I disembarked, blinking in the sunlight, and Wang strode over to me.

"It's a Chinese car, but it has a foreign engine," Wang made sure to tell me, ushering me into his vehicle; even in the dusty corners of China, imports carried status, while domestic brands were something to explain away.

Lauren filed in beside me in the backseat. "Wang will drive us to my massage parlor," she said.

"You own a massage *parlor*?" I exclaimed.

Lauren nodded, glancing at me sideways. Massage parlors were notorious as fronts for prostitution. I doubted she was turning tricks, but it wasn't a good sign she'd leaped headfirst into a business full of competitors who played dirty. "Business is slow," she said, explaining that she'd emptied several years' worth of savings—the equivalent of about $10,000—into buying the venture.

We passed shells of buildings in various stages of construction. Structures of steel and concrete gaped with black holes where windows were to be installed. Cranes dotted the sky; it felt as if we were driving through a Legoland set. The county government wanted to double the population, drawing from the peasant poor of the surrounding countryside. For officials here, as everywhere in China, promotions were earned based on GDP (gross domestic product) growth, which meant that the mantra for development often was "Build it, and hope they come."

"Jingxian government has big plans," Lauren said. The county was a talc exporter until recently, when the local government discerned that setting up processing plants would keep more money in the region. The average annual income of its residents was the equivalent of a few thousand dollars a year, enough to kick it off China's list of "poor counties" a few years ago. It was a typical regional center in China, and there was nothing noteworthy about it, just as there was nothing unusual about Lauren's story. There were hundreds of Jingxian counties in China, struggling to sustain a healthy pace of development, and millions of rural Chinese just like Lauren, working to educate their only child. I pondered the empty buildings, vacant sidewalks, and newly poured roads. Besides our Baojun, there were few cars on the road.

The people hadn't come yet.

"We're just a few months away from *zhongkao*," Lauren said, handing me a newspaper from Jun Jun's school. Printed in red and black on six pages of broadsheet, the paper was teeming with references to student rankings, loyalty to country and Party, and the upcoming *zhongkao*.

"*Zhongkao* is one hundred days away," proclaimed the article at the top of the page, teeming with phrases that seemed like a call to war. "We cannot waste a single minute or second in our dash to the goal. The opportunity to prove ourselves is here. We need to work very hard these hundred days, in order to repay our parents, repay our teachers, repay the school, and give ourselves a better future."

On the second page there was a list of top scorers on the school-wide final exam. "Here we post their names in praise," the paper read, listing ninety-three names divided by grade level. A student essay about humility promoted collective learning: "My friend ran very fast during the 1,000-m run. Always find someone better to motivate yourself." Then, a piece about duty to country: "We need to take responsibility for ourselves, our friends, our country, and society."

The paper was chock-full of propaganda, cobbled together to create both a sense of urgency and a feeling that scoring well was a student's duty to community and country.

"How is Little Jun doing?" I asked Lauren, who sighed.

"He's abandoned hope—*fangqi*—a little bit. We push him to study but he doesn't care anymore."

"Well, he still has time—one hundred days, to be exact," I said, patting the newspaper.

"It's too late. He's very tired," Lauren said. "He says he prefers America's education system. From kindergarten up until high school the parents don't have to manage the children. And the knowledge they learn is useful rather than something you learn only for a test."

"Where did he hear about America?" I asked.

"Movies, television. He said that the American kids always seem so happy."

A headmaster's speech had been printed on the last page of the school paper, word for word: "We the leaders want you to fulfill your dreams—and our dreams. *Zhongkao* is very important and it will decide your fate and your future. Now is the critical moment: Diligence is the only way to fulfill your goal."

We arrived at Lauren's massage parlor, where diligence was nowhere to be seen. Jun Jun sat behind the front desk, engaged in rapid-fire clicking. He was playing League of Legends, a multiplayer battle game involving marauding bandits, monsters, and guns.

"I don't know how to help Jun Jun," Lauren told me.

I glanced at her. I wanted to suggest that Jun Jun bend his head over a math textbook rather than digital monsters, but I kept quiet.

"I've been too busy managing the massage parlor," Lauren said.

I looked around: Not a customer in sight, only a heap of candy-colored six-inch-high pumps. Lauren selected a pair, stepping out of her street shoes and into a pair of green-and-white heels.

Their situation was turning out to be more challenging than I'd thought. Lauren was illiterate and couldn't help with the boy's study. The family's relationship with the boy's boarding mistress—who was supposed to help with test prep—had been severed. They'd depleted their savings to buy this business along with the Baojun-with-a-foreign-engine, and they had little money for tutors. After she'd bought the business, Lauren discovered that the previous owner had been selling sex out of the upstairs rooms. Lauren refused to *qiaodabei*, or "Knock Big Back," so the parlor's regulars stopped coming around and Lauren's masseuses fled for a rival as business dwindled.

Lauren and Wang had arrived back in the countryside, fifteen years too late, to oversee the studies of a boy they'd lost long ago.

⋆ ⋆ ⋆

THE ODDS ARE loaded against rural children in China.

The prospects of a student like Jun Jun are leagues behind that of a Darcy or an Amanda, who attend good high schools in urban centers with resources and access to teacher talent. *Fangqi*—giving up hope—would almost certainly be part of Jun Jun's future. This makes the boy one face of an inequality index that development experts call an urban-rural gap in health, wealth, and education of more than a hundred years.

"I have traveled across more than twenty countries all over the world, and not a single one of them has as big a gap as China," says education professor Zhou Nianli.

As incomes were skyrocketing in major urban centers such as

Shanghai—fueling the staggering media headlines about China's economic juggernaut over the last few decades—the "inequality index" in China rose more than that of any country in the world. China's economic roller coaster had left many families without a ticket, watching from outside the amusement park.

Many of those households are in rural China.

Their children face a litany of problems, starting from inside the womb. A research team out of Stanford University surveyed 350 villages and found that more than half of the babies they tested were malnourished (80 percent, if they included children who were borderline anemic). Three out of four showed delayed cognitive skills. Parents who must migrate for work might be absent for years, like Lauren was, forced to leave kids with grandparents even less educated than they are. By middle school, the researchers found, almost half of rural Chinese children have an IQ *lower* than 90 (average).

All of this is overlaid with a system that is fast-tracking only the 20 to 25 percent of kids who will actually go on to high school. "The teachers ignore the rest of them," says Scott Rozelle, of Stanford University's Rural Education Action Program.

Without the financial, emotional, or academic support to stay in school, these kids fall behind. They're one entrance exam failure away from dropping out of school, and drop out of the system they do, at astonishing rates. Many teenagers go out to work immediately doing unskilled labor—but those jobs are increasingly moving to lower-wage countries such as Vietnam, Bangladesh, and Indonesia. Others languish at home, the welfare recipients of parents who send money home from faraway cities. Some might enter vocational high schools of varying quality.

This fate of uncertainty befalls too many rural Chinese children. When these kids become working-age adults within a couple of decades, they'll be among hundreds of millions of Chinese who won't have the skills to fully participate in a high-wage innovation-based economy, estimate the experts. That's roughly a third of the population.

And, as Rozelle says, "China still isn't rich enough to support a welfare state, so the stage is set for massive unrest, organized and unorganized crime, gang activities, and more."

So here was a reason China didn't allow rural provinces to count in the first two rounds of PISA scores released to the world: There simply wouldn't be much to boast about. (In the most recent round, in 2015, indeed China slipped after Beijing and two other provinces were included with Shanghai's results.)

* * *

I MET LITTLE BAI and Little Cong deep in Henan province.

I was traveling with a volunteer organization that focused on migrant and rural education, and during a free morning I found these girls slurping down breakfast on a street corner teeming with vendors. We were in Luyi County, hours from Shanghai by train-then-bus, due northwest of Shanghai. On Rainey's map of China, I'd be inside the rooster's chest, around where its wing might originate.

"You don't drink soup?" Little Bai asked me. Henan was one of the poorest, most densely populated provinces in China, with an average of 1,464 people packed into every square mile. Little Bai, a slender girl poured into jeans despite the heat, sat on a low stool, before a metal vat of glutinous, maroon-colored broth. The soup sold for three kuai a bowl, or about 40 cents. The soup emitted slow, lazy bubbles with the consistency of boiling paste. A weathered vendor hovered over the vat, and the entire affair was situated a few paces from a bustling street choked with scooters, vehicles, and exhaust.

"No," I said, after considering the question. "I don't drink soup."

"I eat this every day," Little Bai said, bending over her red paste, "so I won't be thirsty."

The girls were sixteen and seventeen, and neither had seen her family in months. Little Cong's parents were nomadic shepherds working in Xinjiang, and Little Bai's parents ran a pharmacy in another town. Their parents had dropped them off in this dusty, nowhere town at a middle school where they could eat, sleep, and study in the same building.

For these girls, the National High School Entrance Exam was eighteen days away.

"Come back to the classroom with us," Little Cong said, slipping her hand into mine as we walked down a long stretch of street. We passed a tree that earlier that day had provided shade for a man in a blue jumpsuit, pleasuring himself during a break from work. "Oh," I'd uttered as I spotted him, and his eyes met mine, expressionless. I'd hurried past the man as his hand continued to pump below the waist; he was the poster boy for a migrant laborer separated from his family by long, isolated workweeks on a construction site. This town was all ambition and no heart, with empty buildings in various stages of construction and few inhabitants save the building workers and the students.

"This school is very decrepit," Little Bai said cheerily, as we approached a coral-colored four-story building, set back from the street and enclosed by a ten-foot-high steel gate. We climbed the stairs to a second-floor classroom encased in dull white plaster. Newspapers dating back two years papered over a row of high windows, letting in only a hint of the sun outside.

"She's an American teacher visiting from Shanghai," Little Bai told her classmates, who sat at desks arranged in four rows. Each desk was covered with books placed upright, forming a barricade between teacher and student that conjured up a Great Wall of Books.

"I'm the fat one," announced a plump teenager, cackling at her own candor.

"I'm the tall one," proclaimed a thin girl sitting in the second row.

"And I'm the stupid one," Little Cong chimed in, to more laughter.

"Who's the best student in the class?" I asked. I was always intrigued when I posed this question to a classroom, as everyone always knew the answer. Ten hands pointed to a skinny girl in black eyeglasses sitting in the front row.

"And where are you?" I asked the Fat Girl, who wore dark sunglasses despite the already-dim classroom.

Fat Girl lowered her hand to the ground, as if she were stabbing at the void beneath the bottom rung of a ladder. "I'm here—low, low, low."

I slipped into a seat between Fat Girl and Little Cong in the third row, and a stern-looking woman sauntered into the room. The class suddenly quieted. The teacher immediately launched into her lesson.

"Let's turn to question eighteen," the woman said. "A truck transported some goods from Place A to Place B. X is the time, and Y is the distance from the truck to A. Use the information in the graph to solve the questions below."

As heads dropped over desks in the front row, I focused on faces and body language, and quickly discerned that the classroom was arranged according to aptitude and ambition. Students in the front row were pert and alert, captained by the No. 1 student, but performance and attentiveness leveled off as you moved to the back of the room. In the second row, Fat Girl didn't even pretend to be interested in solving distances; she was busy whittling down the fingernail on her pointer finger with a pair of scissors, shearing it into a collection of planes and ridges. Two girls in row three took turns massaging each other's thighs.

"Now, write the functional expression between X and Y during the truck's return to A," the teacher said, seemingly to no one. In the back row, three boys played video games on their mobile phones, hidden behind their Great Wall of Books.

It was a long, slow day in June, and the heat and light reverberated in the room as the teacher droned on. To me it was clear: School was a joke to most of these students, since many were just passing time. It was eighteen days until *zhongkao,* and instead of a sense of urgency, I found complacency and acceptance. For years they'd taken practice tests, and few students would be surprised by the outcome.

I whispered to Little Cong, the daughter of nomads, who was smacking gum. "Show me your monthly test papers."

She collected heaps of papers from her desk and crumpled them in her right fist. "Here, take them," she said, thrusting them before me. "I have no use for them."

I uncrinkled one. It was a physics test, and she'd scored four out of fifty possible points. I flattened the others, and they were no better: Eight of eighty points in algebra. In chemistry, she'd scribbled a few characters underneath the first question, but the rest of the paper was blank.

She grabbed the wad of papers and shoved them into my bag. "Take them with you. Look at them later," Little Cong whispered.

A stick of gum made its way to me from row four. "This is from America," Little Cong whispered, as I glanced at the stick of Wrigley's spearmint gum. The wrapper displayed a monkey balancing an assortment of tropical fruits.

"What do you want to do when you grow up?" I whispered to the girls in my row. Fat Girl had propped a mirror against her books and was checking her face, head lowered to escape detection.

"I want to sell fabrics," she said. "I want to put together buyers and sellers."

"I want to be a nurse," said Little Bai.

"I want to manage a hotel," whispered Little Cong.

Over their voices, the teacher said, flatly, "Now let's first look at the graph of this function. What does the X-axis mean?"

In the third row, one girl picked wax out of her neighbor's ear with the pointy tip of a pencil.

The kids had few adult role models, and teachers had become paramount to their development; yet the teachers were often migrants-for-hire themselves, shuttling in and out of classrooms, with little connection to the students in their charge. After another half hour, class was over, and the teacher scooted from the room.

"Where's the teacher from?" I asked, directing my question at row two.

"She's from Shangqiu," Fat Girl said, naming a larger town an hour-long train ride away. "She's only been here for a few months. They never stay long."

"Why not?" I asked. The teacher hadn't acknowledged the presence of a stranger in her classroom, much less inquired why I was there, and now I understood why. Her connection to these children and investment in their future were little more than a paycheck.

"Because we're stupid," chimed in a boy from the back row, finally lifting his head from his mobile phone.

"We only have eighteen more days of school anyways," said Fat Girl. "Eighteen more days, then no more school, ever!"

"Won't you take the *zhongkao*?" I asked her.

"Yes, but I will fail. Then it's no more school!"

★ ★ ★

MONTHS LATER, BACK home in Shanghai, I would glance at a snapshot I'd taken of Little Bai and Little Cong on my last day with them.

The daughter of pharmacists and the daughter of nomads stand arm in arm before a long stretch of empty, unpaved dirt road, flanked by shells of buildings. For a year they'd lived, slept, and studied side by side, but after the *zhongkao* their futures would diverge sharply.

Little Cong, the daughter of nomads, scored a 291 out of 600 points. It was a dismal result, and her parents had no money for alternatives. Cong's family wouldn't be able to buy the several hundred points needed to go on to high school, bribe a principal, or pay a middleman a school introduction fee. In rural areas you could find back doors into academic high schools, but you still needed to score above a certain threshold, otherwise that work-around would be largely out of reach.

Little Cong would go straight to work.

Little Bai, the daughter of a pharmacist, scored much higher: 430. Still 70 points shy of the cutoff to go to an academic high school that year, the score was close enough to pave the way for entrance. Her parents had money, and they were willing to spend. (Such back doors only served to widen the inequality gap; those families with money have options that are life-changing for their children.)

My snapshot was eerily prescient. Little Cong stood thin and lean with her heart-shaped face, crinkled under the sun, her eyes closed when the shutter had snapped. Little Bai gazed into the camera, head high and smile confident, left hand up in a victory sign.

Two years later, I would check in with Little Bai, now mired in study at an academic high school. By that time, her former classmate Little Cong had cycled through as many low-wage jobs as there were years in her life, before finally heading to Xinjiang to join her shepherd parents. Almost three years after she failed *zhongkao*, Little Cong would send me pictures of her heavily pancaked, thickly rouged face, set against the bright light of the massage parlor where she worked. I wondered whether she felt pressure to turn tricks, and I felt sad about her state of affairs.

Little Bai, knee deep in books, was confident she'd test into university, but she sobered whenever I asked about her old friend. If a WeChat message could issue a sigh, hers would:

"Ah, Little Cong. Life is unfair. Society is cruel."

⋆ ⋆ ⋆

I CAME BACK from these rural provinces with a firm understanding: It wouldn't be much of a reach to say that Rainey is part of a giant education experiment.

Consider that just six decades ago, four of every five Chinese couldn't read. The year was 1949 and Mao Zedong and his armies marched against the educated elite and the moneyed—including families like my maternal grandfather's, who owned the central bank in the town of Shanhaiguan. It was an inconvenient time to be wealthy and literate, and Mao and his Communists chased the ruling Kuomintang out of China and across the Straits into Taiwan.

With Mao at its helm, the People's Republic of China was established, and its new leaders quickly got to work. The Communist Party drafted a constitution based on the Soviet model and set about seizing Kuomintang schools and starting new ones in the name of "state-building."

Mao's education officials could only start from scratch: textbooks, teaching methods, basic goals. Qualified teachers had to be hired, a primary school curriculum developed. From there the system evolved throughout a number of his campaigns; Mao's industrialization efforts during the four-year Great Leap Forward starting in 1958 tied education to labor and production goals, while the ensuing Cultural Revolution's halt of formal education saw books burned, teachers humiliated and a complete halt to college admissions during the ten-year period starting 1966.

Since the 1980s, the Chinese government sought to modernize schools and enroll most children in nine years of free compulsory schooling—an era that gave rise to the "quality education" reforms. Leaders have also worked to develop a citizenry strong in science and technology, and those goals, too, were disseminated through the education system. New objectives have been inked in recent years, such as the hope of enrolling all children in preschool and ensuring that nine of every ten Chinese attends senior high school. "My dream is to ensure that we can teach students in accordance with their aptitudes, provide education for all people without discrimination, and cultivate every person in this nation to become a talent," said Education Minister Yuan Guiren in 2013.

As I absorbed the history of modern education in China, I grew dizzy from the flips and turns. Take the attitude toward educated intellectuals: Depending on the decade, they were alternately praised and emulated, reviled, banished, or killed depending on their views. Education in China has always been characterized by "bold moves, major shifts, and reversals," as the researcher Mun Tsang wrote.

Still, there was no mistaking how much progress the country had made in six decades. *Putonghua*—the Mandarin most educated Chinese speak today—was only adopted as the national dialect in 1956.

More than a few researchers have dubbed the Party's efforts to eradicate illiteracy "perhaps the single greatest educational effort in human history." In the following decades, the Chinese government established a school system that educates a fifth of the world's population, and with basic literacy and compulsory education established, the government would continue to refine its goals.

Change would begin in cities such as Shanghai, where funding and connections and the openness of city folk mix and mingle in a laboratory of sorts. Schools that lacked access or the right confluence of circumstances, unfortunately, would lag.

There isn't enough money for "everything all at once," said Wang Feng, my Ministry of Education source. "We're not a developed country whose wealth is sufficient to finance every school toward the right track of development. So our resources are distributed first to develop some schools, while the other schools carry out the duty of reducing illiteracy. This generates a gap between good schools and bad ones."

Stanford researcher Scott Rozelle laments that education in modern China suffers as Beijing focuses attention elsewhere: Divert a mere "ounce" of the money that China devotes to sending a man to the moon, building a Xi'an-to-Urumqi high-speed rail, or giving to Africa, he says, "and it would change the face of rural China. It is not about having enough money, it is about priorities."

Whatever the case, progress is not evenly distributed. The quality of Little Cong and Little Bai's education was only as good as the teachers who cycled in and out of their classrooms, and in rural areas the teachers changed stations as frequently as the trains at the nearest depot (as the government encouraged migration to urban centers, rural schools were also consolidating and enrollment was dropping). Jun Jun's school in Anhui province, and even Little Pumpkin's school in Shanghai, notched far below Rainey's on the educational quality totem pole.

Going to school in Shanghai offered a student other advantages. The city was also a province, making Shanghai a special education zone with the autonomy to write its own curriculum and launch special initiatives.

For every child privileged enough to wrestle his or her way through schools in Shanghai, there were a handful of children in the countryside who were struggling without proper support in poorly resourced schools or had been left to fend for themselves educationally as their parents toiled in another city. These factors would combine to form a stubborn inequality that would continue to be a drag on the system.

As long as the Jun Juns existed, the educational triumphs of teenagers like Amanda would come with a bittersweet footnote. That note might read something like the words of Suzanne Pepper, who wrote that education can be a great "destabilizer . . . given its capacity to raise aspirations faster than developing political institutions could satisfy them."

There would never be true progress as long as rural education continued to struggle.

* * *

OUTSIDE JUN JUN'S bedroom window lay his first job opportunity.

"A construction site! If you fail *zhongkao*, you can go straight to work," his father teased his son.

Jun Jun snorted from his desk, piled high with books, pausing to ponder his view of workers who transported earth between gaping holes and tall piles all day long.

I'd come back to see Jun Jun and his family the week before *zhongkao*. His father, Wang, collected me up from the long-distance bus station in his Baojun-with-the-foreign-engine, and drove me to their rented apartment. The floors were unfinished, doors still wrapped in blue manufacturer's plastic. In Jun Jun's bedroom, the three of us were close enough to smell each other's breath.

"This will be on the *zhongkao*," Jun Jun muttered as Wang settled into a plastic chair near his desk. I sat on the bed, where I could see the boy's hunched back. "This will be on *zhongkao*, too. This will not, so I won't need to study this." The boy stooped over his physics book, flipping pages, pencil scratching on paper, following by long periods of silence. He seemed distressed, and he muttered to his father in their Anhui dialect.

"What's the matter?" I asked Wang in Mandarin. Wang's foot tapped the concrete floor.

"He doesn't understand any of this homework," Wang whispered. I stood up and spotted a math test on the desk. He'd received 102 out of 150 points.

"That's not bad," I muttered encouragingly to Wang.

"It's not enough to pass *zhongkao*," Wang said.

"I can't even write out this equation," Jun Jun said, swigging from a glass jug of green tea. The tea leaves drifted in the beige-yellow liquid, aimless.

Jun Jun was now working on a calculus problem set involving derivative proofs. China's curriculum was famously difficult, and content in heavily populated provinces such as Anhui was particularly advanced, so that it could effectively weed out large numbers of students.

"Keep trying," Wang urged Jun Jun.

"Do you sit here every night?" I whispered to Wang.

"Yes. I don't understand any of the papers, but I like to make sure he writes something on every page," Wang said, lowering his head. "His education level is much higher than mine."

Jun Jun could read and speak some English, he was testing well on advanced math, and he was already touching upon calculus. He'd already attained what any Westerner would consider a high level of education for a sixteen-year-old; but in rural China this wasn't always enough to get into an academic high school.

"What will he do next year if he doesn't pass *zhongkao*?"

"I don't know," Wang sighed. "He's so young. I don't know."

But Wang knew, even if he wouldn't say. The construction worker wasn't good with books, but he was wise in the ways of modern China, and he knew Jun Jun would eventually follow the same path his father had twenty years earlier, out to the factories and building sites. Though, I wondered, as China's economy slowed and its manufacturing base inched over to other parts of Asia, whether Jun Jun would be able to find work as easily as his father always had.

It was nine p.m. on a school night. Their son was on his last stop in China's nine years of compulsory education. The family had staked its entire working life on financing him—with textbooks, test prep, tutoring, and boarding homes—through to the next rung of education, but it appeared Jun Jun was about to fall off the ladder. Lauren and Wang tried to make up for fifteen years of missed parenting, but *zhongkao* was marching toward them full steam.

Lauren remembers journeying home at Chinese New Year to visit her little third-grader. Her prescient nine-year-old revealed a hatred not only for his name but for the path that it signified.

"*Jun Jun*. Soldier—that's a stupid name," the boy had said. "Who wants to be a soldier anymore?"

9

SHORTCUTS AND FAVORS

If you have money, you can get ghosts to do your work.

—CHINESE PROVERB

Teacher Song had a proposition for me one spring afternoon, though I didn't immediately recognize the offer behind her words.

"Rainey's attention span needs fixing," she told me at pickup, as children weaved around us on the way out of the classroom. She had a habit of standing in fifth position, like a proper ballerina, heels locked together and toes pointing toward opposite walls.

"*Buhaoyisi*—what do you mean?" I said, startled. We'd settled into an easy rapport buttressed by head nods and waves, but Teacher Song didn't usually address me so directly.

"His attention span is poor," Song told me. "You have to practice at home with things that require concentration, like puzzles."

Earlier in the year, Song had sent an article via WeChat about cultivating "focus and attention." "That is the important, first step toward genius," the article declared:

> Subject your child to the stare-point method, by requiring him to fixate on a single spot for several minutes daily. Upon success,

deliver positive reinforcement with verbal praise, touching and kissing. Then the child should rest for 5 to 10 minutes.

I was decidedly *not* a stare-point practitioner, and apparently Rainey had suffered. "He has problems of concentration. You need to practice with him," Song barked suddenly, probably noticing that Mom had lost her train of thought, too.

The following week, Song delivered the second part of a one-two punch: a chart documenting every child's progress on the recorder. The thing came with a disclaimer: "Don't compare your kid to other kids. This is just a report to let parents know how well your child is doing."

In private, I seethed. She'd sent the chart to all the parents via WeChat. How could we avoid comparison with other children when scores are presented side by side?

"How's Rainey doing with recorder play?" she asked, finding me the next day.

"Well, as you said," I told her pointedly, "he has trouble with rhythm."

"When I pay attention to Rainey, he plays certain notes. He's not bad when he *concentrates*," Song said, enunciating this last word. She glanced behind me to find an empty hallway and focused back on me. "Would you like me to spend some extra time with him?"

Immediately I understood.

"Oh!" I said, shifting my weight, trying to decipher what this meant. Extra time during class? Or outside of class, on evenings and weekends? And, do I give her money for this "extra" time? What if I offer money, and she's offended?

Worse, what if I *don't* offer money, and she's offended?

Just before the beginning of the school year, teachers were officially banned from accepting gifts and money, and from offering tutoring-for-hire to their students outside of class. Straight out of the Ministry of Education, this 2014 prohibition was part of an ongoing effort to root corruption out of the education system. Soong Qing Ling issued its own interpretation:

> Adhering to the teaching principle of Founding Grandmother Soong Qing Ling, all the teachers will treat every child equally and refuse any form of "gifts" from the parents. Our committee appeals to the parents to give up giving gifts. Let's create a truly harmonious environment for children to develop.

Inside of this newly articulated anticorruption environment, I realized that Song's query had special meaning: It signaled acceptance into Middle Class No. 4's circle of trust. Foreigners were rarely offered an invite, but clearly Song thought of me as a Chinese face.

Through the gossip grapevine that kept tabs on teachers, I knew that certain teachers at the school accepted cash and gifts from families deemed "safe." Fine imported wines. A box of French hand creams. A gift card loaded with 10,000 RMB—that was $1,700, or a couple of months of a Chinese teacher's salary.

Stepping inside that door meant entering an illicit world of gifting for favors, gifting for a teacher's attention, gifting for grades. Once you pass through, you cannot easily turn back, and the student-teacher relationship is forever altered.

"Uh," I stuttered, glancing at Teacher Song, "I—let me think about that and get back to you."

Song nodded, and I backed away from her.

Once I rounded the hallway corner and was safely out of sight, I ran.

* * *

THE CHINESE PEPPER their speech with the inspirational proverbs of ancient wise men, yet the saying I found most appropriate for Chinese society reads less like a motivational saying. To me, it is a condemnation:

> 有钱能使鬼 推磨: If you have money, you can get ghosts to do your work.

In other words, "the rich can awaken the dead," as another interpretation goes. Money makes anything possible—it can even rouse the spirits from slumber.

Gifting has long wielded immense power in Chinese society—I certainly felt the pressure early on in Rainey's journey—and this quality, coupled with today's runaway consumer culture, has made such a greasing of the palms a part of everyday life. Reciprocity is almost always expected. "Someone gave me a peach as a gift, so I sent him a pear in return," trumpets the *Classic of Poetry*, one of five classics of the Confucian canon. "It's improper not to return what one receives," proclaims another famous proverb.

Gifting inside a schoolchild's journey might start innocuously, like a pineapple cake to a principal or teacher, which he or she graciously accepts. It's just a token of appreciation, but, nevertheless, a microscopic line has been crossed. Then, you hear Nong Nong's mother delivered cash in a red envelope, and you notice shortly thereafter that the teacher gave the boy a front-row seat in math class. (Your boy sits in back, where it's harder to hear.) Soon, you find yourself shopping for Louis Vuitton at Chinese New Year. Within a few months, you hear jobs are the newest gift and that Mei's father secured an internship at his pharmaceutical company for the school principal's college-age niece. Meanwhile, your son gets "elected" class monitor—in China, the word "elect" must always be book-ended with quotes—and you wonder whether that had anything to do with the Louis Vuitton.

The desire to gain an advantage, any advantage, settles deep into the pit of the Chinese stomach, and before long you're loading up gift cards with renminbi and slipping them into the teacher's palm. What started out as a pineapple cake has suddenly become, "I just gave you five thousand RMB—What will this do for my daughter?"

This system of gifting and reciprocity favors well-funded parents, of course, whose children might be granted a fast pass onto the highway of individual attention and opportunity. Rewards might include extra test prep, special awards and leadership positions, opportunities to head overseas for school exchange programs, and even back-door admissions to high school and college. Parents who cannot afford to participate, or who refuse on principle, may see their children struggle.

Amanda and her parents learned their lesson early. "Teacher Tang, Teacher Tang," Amanda suddenly blurted out once, rapping her knuckles on the table as we talked—a sudden movement for such a slight girl.

"Who was Teacher Tang?"

"My elementary school *banzhuren*—homeroom teacher," she said, as a shudder passed through her shoulders. An eight-year-old Amanda was terrified of her. "Teacher Tang always belittled me in front of my classmates. She'd say, 'You may think you're smart, but you're just normal.' I didn't know why she was picking on me, but we were taught teachers were always right. I thought I was a bad kid."

The verbal tirades worsened when Amanda performed well, when she sang with a particularly clear voice or snagged first place on a math test. When Amanda gathered the most "votes" for class monitor, "Teacher Tang would rig the election so that I would come in second. I never 'won,'" Amanda says.

Years later, Amanda could still recall the Teacher Tang stare: so hot and intense that she rarely saw the whites of her owlish eyes, only dark, gawking black pupils.

"Why did she do this to you?" I asked.

"We never paid," Amanda said. Much later, another student confessed to her that her parents had been plying Teacher Tang with gifts.

"Like bribes?" I asked.

"Gifts. Giving her money. Paying for trips. One student in our class—her family took Teacher Tang on vacation to Hainan Island." The tropical island was China's version of Hawaii. Most of her classmates, Amanda estimated, had been giving the equivalent of $100 to $200 a few times each year, which amounted to a significant sum for a middle-class family.

"Later, did your parents regret not participating? Not playing the game?" I asked.

Amanda stared into her coffee. She nodded once, quickly, almost imperceptibly.

Another time, Amanda found trouble at the high school registrar's office. She needed a copy of her transcript to apply for the US high school exchange, and she'd approached the man in charge of grade reports.

"I'd gotten As in everything—I always got As. But there were Bs all over my transcript," Amanda told me, her voice quivering in anger. The man had handed her a transcript with her name emblazoned across the top, and Bs listed below in politics, Chinese, and biology.

Flummoxed, her parents asked around, before a friend finally clued them in. "Don't you know? You're supposed to pay," he'd told them.

"The going rate was two thousand RMB," Amanda said bitterly, bringing me before-and-after copies of her transcripts when I pressed her for proof. I noted identical documents, down to the red-inked chop of the school, *except for* the letter grades in the three subjects of politics, Chinese, and biology.

"After my mom paid him the money, he gave me back the As," Amanda said, tapping the second paper. "There's a basic amount you have to pay to get your proper transcript, and the number increases every year."

"What happens if you don't have the money?" I asked.

"If I didn't pay, then I would just get the three Bs on my transcript."

"Teachers shouldn't be accepting gifts," I exclaimed to Amanda. "The registrar shouldn't wield his power to get cash. That's a conflict of interest."

Upon Amanda's return from her year at a US high school, a litany of observations about American culture had tumbled out of her, insightful in their simplicity: Western parents want achievement, but they don't want their kids to suffer for it. Self-esteem is very important. So is football. Rich people have the connections to continue to be rich and make sure their kids are rich, too. Yet in her time there, Amanda hadn't often come across the term "conflict of interest."

I struggled to explain. "Should Teacher Tang have been accepting gifts?" I asked Amanda. "Families who give gifts may enjoy unfair advantages."

"The Chinese don't see it this way," Amanda says, with a shrug. "This kind of behavior is everywhere."

My migrant friend Lauren also fell victim to such schemes: Jun Jun's teachers wanted money for school newspapers that should be free, fees for off-syllabus books, cash for outside classes held by homeroom teachers who declared them "mandatory." "Jun Jun's teacher said if he didn't attend, he would fall behind during regular school hours," Lauren bristled. "Eight hundred RMB per month. There is no *shizai*—no honesty."

Students take shortcuts, too. "At least ninety percent of my students cheat," estimated one principal of a Shanghai high school. "Every morning, I'd take a tour of the building and catch at least two students copying homework." Lately, he added, technology has made cheating more difficult to catch. "They use WeChat on their mobile phones. They take pictures of tests and send them to friends."

Darcy revealed an educational back door of his own one humid, rainy afternoon. "My father is not married to my mother," Darcy told me, voice subdued as we walked toward the subway station after a meeting. The revelation was an accident; I'd spotted an inconsistency in statements he'd made about a "stepmother" and asked him about it.

"They're not married to each other?" I asked, unsure where we were headed. Was this a conversation about a parent's midlife crisis and subsequent affair?

"No, they're not married," he said, pulling his umbrella low over his face. He kept his eyes on the sidewalk as we strolled. "When I was ten, my father took vows with another woman. I call the woman my stepmother. They are married in name only."

The stepmother is from Shanghai, he continued, and suddenly

everything made sense. *Married in name only*. His father had brokered an arrangement in secret and greased it with a pile of cash: His father took a new wife, the woman got financial security, and the boy obtained a parent with a Shanghai *hukou*.

Hukou refers to the household registration system that links a child permanently to his family's hometown. Generally, Chinese children attend high school and also sit for the National College Entrance Exam in their hometowns. That means the most critical years of a child's education are governed by what's printed on his or his parents' *hukou*. If there were an American equivalent, I might be forced at age fifteen to relocate to my birthplace of Philadelphia to attend senior high school, even though I'd moved away when I was in diapers and no longer knew anyone in the city.

More than that, entrance exams for college vary in content depending on where you take it, and universities allocate more admissions slots to some provinces. The American equivalent might be SAT tests whose questions were more difficult for students in Omaha than those in New York City. That means a *hukou* could very well bind an unlucky child to a province where entrance exams are more difficult or where fewer kids advance into the top tier of colleges. For example, Tsinghua University—commonly called the Harvard of China—accepted roughly two hundred kids from both the city of Beijing *and* the entire province of Henan in 2016, despite the fact that Henan has seven times more people than Beijing at ninety-five million people. (Shanghai kids benefit from dice that are similarly weighted; students with a Shanghai *hukou* were fifty-three times more likely than the national average to get into the premier Fudan University.) In China, the *hukou* system is a deterrent to social and class mobility, and some researchers have likened it to a caste system or "Chinese apartheid."

Darcy could have been one of those disadvantaged kids. He was born in rural Hubei province, but was transported to Shanghai as a youngster on the heels of a mother and father who sought a big-city future. A decade later, Darcy is an urban teenager with city manners, his rural Hubei dialect long ago swallowed up by his adopted megacity of twenty-six million. Meanwhile, the hometown listed on his *hukou* had become an economic wasteland.

"I saw fields lying fallow and lots of empty houses," Darcy told me, describing a trip back to his village for a relative's wedding. "As a boy I went to my cousin's house and played in the fields—back then they were lush and green." His was a countryside emptied of jobs and people.

The solution was clear, the family decided. Darcy must attend high school where progress hangs its hat. He must sit for *gaokao* in Shanghai.

So his rural *hukou*-holding parents divorced, and Dad fake-married up to someone with an urban *hukou*, for the price of 50,000 RMB. That qualified the boy to stay in Shanghai for schooling.

Skirt the rules and get a better life. It's a compelling incentive, all right.

I found this revelation confounding. "After the *gaokao* is over," I uttered dumbly, in the direction of Darcy's umbrella, "will your mother and father get remarried?"

We reached the subway, where he faithfully escorted me after each meeting. "See you next time," Darcy said, his voice growing faint, as if suddenly realizing he'd revealed too much. I stooped to peer underneath the brim of his umbrella, and he tilted his head forward to say good-bye.

* * *

ON DAYS I wasn't feeling empathetic, I would brand the entire system with a single word: *fubai*—corrupt or rotten. Each time I saw a news report about a principal who'd taken bribes, a teacher enrolling her students in after-school prep for side money, or an elaborate *gaokao* cheating scheme, I found that word popping into my head. *Fubai*.

It was the judgment of a Westerner schooled in a different culture.

Over time, I landed upon a simpler truth, one less judgmental and more reflective of the reality of China today: The rules are so rigid and hierarchical, and the game is zero-sum with incredibly high stakes, so that to survive, the Chinese had become accustomed to seeking a work-around. Breaking the law, or a matter of survival?

A schoolchild's journey is full of assessments and evaluations, which are typically made public—posted on Big Board, say—for all to see. When they're not broadcast formally, results propagate just as rapidly via the invisible scorecard of gossip: Which child raised his hand most often? Whose recorder play most closely followed rhythm? Who snagged first place on the math exam? Whose *gaokao* scores topped the district list?

A score would simply be a score, and nothing more, if such assessments were simply a periodic check on progress. In the Chinese school system, a number is much more than that. A relentless churn of testing coughs up scores with real consequences, and the stakes are many, high, and intricately intertwined: A teacher might get a salary bump for cultivating high-scoring students; a principal might be granted a promotion based on his school-wide *gaokao* average; a student might gain admission to an elite college, while her classmates lose their grip on the ladder to social and economic security. Couple this high-stakes, zero-sum environment with a large population, a scarcity of opportunity, and a cultural propensity to give gifts, and suddenly the education system becomes riddled with *houmen*, or back doors, through which gifts and money exchange hands.

In today's China, it would be tempting to brand as *fubai* the teachers and administrators who dabble in the gray. Certainly, some I've met would easily classify as greedy and spiteful, but most are pawns caught up in a system beyond their influence and design. The notorious pride of the Chinese also plays a role; losing face or admitting defeat simply isn't an option, and taking the moral high ground would mean falling behind. It's a national game of Keeping Up with the Wangs, and sometimes that requires using every tool in the kit.

There's an additional problem: Teacher and administrator salaries are relatively low, which means that opportunities to earn extra cash can be critical to keeping afloat in modern China. Educators may have the reverence of Chinese society, but respect is little consolation when it's the factory owners, entrepreneurs, and professionals who have cashed in on China's economic growth. Teachers also face the unique pressures of an authoritarian system. For one, they must curry favor with superiors who determine promotions based on assessments. A teacher who oversees evaluations at a Shanghai school explained that he might walk into a music or art class, select students at random, and ask them to sing a song or draw an object on command. "If the child knows the song or draws well, the teacher is doing a good job." A math teacher might be evaluated based on the test scores of her kids. These assessments are slotted into a file and cracked open whenever teachers apply for raises and promotions.

This places immense power in the hands of the assessor. Teachers can only hope their judge is fair, in the same way a 1980s Chinese farmer might have longed for a local tax collector who resisted embezzlement or unjust, random levies. Not everyone is so lucky. "It's an open secret that we must offer up to 50,000 RMB to administrators in order to be considered for promotion," wrote one teacher from Ezhou city, in Hubei province, on an online forum.

Amanda told me her high school math teacher was always looking for back doors. The man sat on a committee that formulated exam questions for a citywide math contest, and this power proved to be a temptation he couldn't resist. "He distributed the questions to his own students," Amanda told me, "in advance."

Armed with this illicit head start, Amanda said, the man's students performed among the top in the district. The teacher's own performance evaluation sparkled that year.

"What happened then?" I asked Amanda, imagining the man was found out and fired.

"He got a promotion," Amanda told me, "and the school got a prize."

Chinese media have reported many dozens of cases of large-scale fraud around education over the last few years. A high school principal in Guangdong took "fees" from parents to enroll hundreds of kids with entrance scores that fell short of the cutoff. More than three thousand teachers and administrators in Hubei required students to buy school uniforms priced higher than market, then raked in kickbacks from uniform companies. A rash of administrators in Guizhou province were dismissed for taking bribes; an Anhui administrator purchased scientific equipment and used kickbacks to buy an apartment for his son. A Renmin University official confessed to taking more than $3.6 million in exchange for helping students secure spots at the college or for other favors surrounding the admissions process.

Authorities have tried to root out the problems, only to find that policy is one thing and reality quite another. The system's massive size, and its firmly entrenched education culture make change difficult. Sometimes, ironically, teachers and parents themselves object to anti-corruption efforts, comfortable as they are with a system that is clearly flawed but ultimately familiar.

Take government efforts to root out cheating at *gaokao*. Many shortcuts have been documented for this important exam—crafted by students, parents, and even teachers who benefit when students test well—and they get more clever each year. News reports show photos of microphones hidden inside coins and eyeglasses, receivers shaped like pencil erasers, tiny cheat sheets. Students from one province have been hired to take tests for those in another, a particular problem in rural areas. Consultants can be hired to transmit answers on test day, and paying an official for an advance peek at the test isn't unheard of. Other times, teachers have been known to sell the answer keys.

In recent years, Beijing officials have proposed extreme punishments, such as a 2015 law that penalizes *gaokao* cheaters with a three- to seven-year prison sentence. Local education bureaus are also attempting their own measures, such as a ban on watches in the exam room. Guangdong province instituted an ID recognition system to prevent proxy test-takers from sitting in for students. A Jingzhou school administered the test deep inside a forest, where no cell signal could reach, and a photo that went viral shows desks placed more than an arm's length apart as teacher paced the gaps, presumably to prevent the exchange of answers. The ministry recently decreed that every testing room in the country must be monitored by at least two staffers.

It was a public admission of an epidemic.

Yet, crack down too much, and people get angry: "We want fairness. There is no fairness if you do not let us cheat," chanted a mob of angry parents, after officials unexpectedly installed metal detectors and outside test proctors at a test site in Hubei province. Cheating is a "nationwide pastime," this group of more than two thousand parents claimed, as they threw rocks and chanted at test monitors, who took cover inside the building.

Cracking down only on their children puts them at a disadvantage, they said.

★ ★ ★

AROUND THIS TIME, I returned to Little Pumpkin's classroom.

I sauntered down the long hallway to Small Class No. 1 at Harmony Kindergarten and spotted Master Teacher Wang at a sink in an open restroom. The way she washed her hands brought back all sorts of memories of my first days in her classroom: As she ran her hands under the water, only the butt of the hand and the fingers touched, forming a cavity between her palms as if she cupped a baby mouse. When she rubbed vigorously on those two contact points, it seemed as if she were torturing the tiny rodent with a great amount of determination.

I trailed her into the classroom, and, once inside, we struck up a conversation at the back of the room as the associate teacher launched into a lesson. "Where's Wang Wu Zhe?" I asked Master Wang, scanning the children's faces for Little Pumpkin, anxious to see how the little boy was faring six months after I'd first observed him in the classroom.

"Nnnnh—he's not here. He's sick. He's always sick," Master Wang told me, leaning against a bunk bed. Wang's face was all angles, sharp features settled into a pointed frown.

"Is he having trouble in school?" I asked her.

"He is a *waidiren*—outsider," Wang told me, as if her big-city bigotry would explain a rural boy's absence. "You want to see another strange child? Take this one," she said, abruptly stepping forward to grasp a little girl by the shoulders. "*Zibizheng*—she doesn't understand anything." The three characters in *zibizheng* translated to the words "self," "shut," and "illness." The girl was autistic.

"Her parents refuse to do anything about her," Wang told me. "She'll just sit in the back until one day she'll drop out," Wang concluded. It was a "do-nothing" approach to special needs, and the girl sat near the back of the classroom, emitting noises that formed no words I could recognize. "*Bang hai bao, bang hai bao.*"

Teachers Wang and Li had covered much ground since I'd first visited. The kids had memorized a dozen nursery rhymes, learned to count, and sussed out the differences among planes, trains, and cars. Wang surveyed her ordered classroom with pride—tin water cups lined up to attention in a cupboard, folded pajamas atop each pillow, children perched silently in chairs—as if the occasion of my visit had suddenly clarified her success.

"The children are very *guai*—well behaved," I affirmed.

"They wailed like ghosts and howled like wolves that first week," Wang said, "but now—now they listen to me."

An outlier quickly emerged. As the children worked on a penguin coloring exercise, seated in clusters of eight, little black bodies of flightless birds began to emerge. One little girl inexplicably drew two large, round eyes on the penguin's left cheek; this was a violation of Wang's classroom, which could never be mistaken for an exercise in free-form painting or Picasso-style cubist sketching.

The penguin was supposed to be drawn strictly in profile.

In three steps, Wang bounded over to the girl's table. "What are you doing? You are drawing two eyes on your paper! Two eyes! Look up at me," Wang commanded.

The girl obeyed. Deliberately, Wang turned to profile, to offer the girl a view of her left-facing silhouette.

"When I turn to the side, do you see *one* eye or *two*? ONE eye or TWO?" Wang commanded with that sharp voice that made heads twitch.

The girl parted her lips, but nothing came out.

Wang snapped. "LOOK AT ME! LOOK AT THE SIDE OF MY FACE. DO YOU SEE ONE EYE OR TWO?"

"One," the girl finally uttered.

"That's right. ONE. ONE EYE!" Wang said, rapping the table. "So why are you drawing TWO EYES on your paper?"

★ ★ ★

AFTER LUNCH, Master Teacher Wang moved in on a golden opportunity.

"Your husband is in America? When will he be back?" Wang asked. I'd mentioned that Rob was away on a business trip.

"Next week," I said.

Wang's eyes brightened. "Well, maybe he can bring back something for me," she said. She turned to Teacher Li and conveyed the news of Rob's whereabouts with the jollity of a five-year-old on Christmas morning.

Teacher Li knew immediately what this meant, and she motioned for the children to file into their chairs. Li had a softness about her that made her approachable, with her bouncing pixie cut and an aloof grin. At times, she seemed embarrassed by her partner teacher, especially when Wang screamed at the autistic girl or yelled fiercely at a boy who rose for water out of turn.

On the subject of shopping, the two women were perfectly aligned.

"What purses do you buy?" Teacher Li asked, glancing down at the bag at my feet. My chocolate-brown canvas messenger bag was the functional tote of a writer who carries a laptop wherever she goes.

"Um—I don't really buy purses that much," I stammered. In the eyes of Wang and Li, I had zero style.

Li continued. "We don't use Coach anymore. Too many people

have Coach now," she said, raising an eyebrow. Was she trying to send me a message?

Wang chimed in. "Have you heard of *Qianbi*?"

I sounded out the phonemes in my head, but they strung together to form no brand name I knew. "No," I admitted, reluctantly. Wang motored over to the classroom computer, typed in a few characters, and up popped the image of a familiar yellow bottle of lotion, displayed before a familiar candy-green-colored box.

"*Qianbi!*" Teacher Wang repeated.

"Clinique!" I exclaimed.

"Yes!" Wang grew excited. "What about *Maike Gaoshi?*"

"I don't know *Maike Gaoshi*," I said. A few keyboard strokes, and a purse popped up on the screen, displaying the initials MK.

"Oh! Michael Kors!" I exclaimed.

"*Tangli Baiqi*?" Wang said, loading a circular symbol comprised of Oriental-looking etchings.

"Tory Burch!" I said.

"Yes!" Wang said, pleased I knew this one right away. "That—this is more exclusive. Too many people have Coach now—about six of ten people on the street have Coach now." Was that a scientific study?

We worked our way through the Chinese names for Louis Vuitton, Kate Spade, and Marc Jacobs, and finally Wang pranced over to a metal cabinet at the side of the classroom. I'd thought it held art supplies, but it turned out to be a secret repository of luxury goods. Wang triumphantly threw open the doors, nudged aside a box of markers, and produced a black clutch.

"*Pulada*—this was five thousand RMB," she said, caressing the patent leather. Prada. Next, she excitedly pulled out a mobile phone and showed me a snapshot of various luxury items positioned against a wall, as if a handbag had committed murder and was ready to be fingered in a lineup of purses. "My sister brought all of these back from France. She also bought four *Laolishi*."

"*Laolishi?*"

She tapped on her keyboard again, and an image popped up: Rolex.

"You know," Li said suddenly, pointing in my direction with her nose, "Teacher Chu is American. The American brands will be cheaper in America. We'll buy the European brands in France or Korea."

My marching orders were clear, and I slowly pulled out my MacBook Air to take notes. I realized I'd made my bed by gifting the Coach purses, and I was uncomfortable in my new role, but I also couldn't figure out how to extricate myself from this situation.

The children sat silently in their U-shape, awaiting instruction, as Li's eyes suddenly fixed on my laptop. "How much was that," she said, tapping the glowing Apple logo.

"I bought this in the United States—for about two thousand American dollars," I said.

"When is the new iPad coming out?" Teacher Wang said.

"Can you buy us the iPhone 6?" Li chimed in.

Within five minutes, I found myself in a conference room, six teachers peering over my shoulder. Wang had yelled into several open classroom doors as we marched down the hallway, "Teacher Chu is going to buy me a *Maike Gaoshi* purse!" prompting other teachers to melt into the procession. I'd suddenly become the Pied Piper of luxury goods. The foreigner who'd been, at best, an intrusion in the classroom, suddenly had value: China's tax system made foreign-branded items purchased inside China prohibitively expensive. Apple computers might cost a third more, while a Louis Vuitton handbag might be double the price of one purchased in the United States or Europe.

This time, Rob would be their mule. I felt the women's breath hot over my shoulder as I toggled between the sites of Nordstrom, Bloomingdale's, Macy's, and Zappos. Where is the principal? Who's watching the children?

Li wanted a navy blue Michael Kors satchel in a style called the Selma. But only Nordstrom had it, in green.

"I can't find it," I said, reluctantly.

"I really . . . want . . . that bag," Li stated, voice staccato, breath landing on my cheek. "Not in green. In blue."

"That blue is last season's, and they're gone," I concluded.

"Will they make more? Call them," Wang ordered, hitting me on the shoulder.

Call Michael Kors? "Let me look again," I said, nervously. My fingers slipped over the keys, and Nordstrom's website popped up again.

Wang struck me again, this time on the forearm. "You already looked there," she said.

"I'm sorry," I said, glancing up at her like a chastened schoolgirl.

"Forget about it," she responded.

After another fifteen minutes, I had my orders. Four teachers settled for Michael Kors wallets, while Wang chose a striped Tory Burch straw bag that cost roughly a month's teacher salary.

"I like a straw bag, especially in the summertime," Wang said conclusively, while the women murmured appreciatively. Two generations ago, tens of millions of Chinese starved to death in the countryside due to one of Mao Zedong's failed policies. Now we were surfing for handbags and matching purse styles to seasons of the year. Starvation to luxury bags in fifty years: If anything encapsulated the speed—and the irony—of China's transformation, this was it.

I glanced up at the gaggle of teachers huddled over me, pondering how to phrase my next question. "How can you afford all these purses?" I finally asked. The average Shanghai schoolteacher's salary that year was about 750 American dollars a month, just about the cost of a Tory Burch handbag.

Wang redirected. "Why don't you buy purses?"

"I have two boys to raise," I said.

Wang nodded. "Do you have to buy apartments for the boys so they can get married?"

"We don't have that custom, but some American universities cost up to three hundred thousand RMB a year," I said.

"*Wah!* We only have one child each, and college tuition is at most twenty thousand RMB a year," Wang said triumphantly. "I already have an apartment, and I have a daughter, not a son. So we can buy lots of purses!"

I loaded the goods into my online shopping cart and typed in the New Jersey address where Rob was staying. "Finished!" I declared.

As a group, we made our way back to the children, teacher by teacher peeling off into corresponding classroom as we moved down the corridor. When Wang and Li and I reached the end, we found our children seated quietly in their classroom, huddled over their water cups, the classroom ayi keeping watch.

"Let's start Three Little Pigs!" Teacher Wang proclaimed, stepping to the front of the room, launching this British fairy tale in Chinese rhyme. Educators in China loved this story, as its moral mirrored Chinese beliefs about hard work. Why erect a hut of mud or twigs when you can invest the effort to build an indestructible house of stone?

I took a spot at the back of the room. I found my body relaxing into the chair—ironically, I felt more welcome than ever before—as my mind raced over what had come to pass. The Coach goods offered at my first visit had instigated a brazen, greedy frenzy of luxury purse shopping on my second visit, and I simply didn't feel I could refuse. These teachers might use those Michael Kors wallets to satisfy favors owed in their own lives; meanwhile, the gray areas of the system would continue to throb and thrive, fed by the players trapped within its margins.

Up front, Teachers Li and Wang launched their lesson with unusually good cheer, probably buoyed by thoughts of retail bounty.

"The pig voice should be light and fast, and the wolf voice should be deep and scary," boomed Teacher Li.

Li blew heartily as twenty-seven children followed suit, swaying to and fro from their seats, and Teacher Wang didn't yell again at the children for the rest of the hour.

* * *

MY HUSBAND WAS a reluctant mule for Tory Burch.

"This thing is bigger than my carry-on!" Rob exclaimed, calling me from his brother's home in Princeton, New Jersey. "You want me to bring this back to China?"

Rob texted me a photo of the package. Ding! The straw bag of the 2014 Tory Burch collection was enveloped in orange cardboard and secured with a gold seal, and the entire arrangement was the size of a baby elephant.

"Yes, it's big," I admitted, "but I need it."

After some heated discussion and a few minutes of cajoling, my harried husband finally agreed to drive to the nearest Home Depot, purchase a shipping box, and bag-check the monstrosity for the long flight back to China.

The evening of Rob's arrival, Wang pulled into our complex in a Volkswagen SUV. She'd warned me against coming to school. "We have a new principal now, and she has an office inside the school. I'll come to you. I'll bring cash."

The sky was dark. I stood under a streetlamp, at the trunk of Wang's vehicle, adopting the hunched shoulders of a dealer ready to palm cash for drugs, Tory Burch bounty in hand. A shadow emerged from the driver's seat and made her way toward the back of the Volkswagen, which, like luxury bags, was also subject to formidable import taxes. Wang was a teacher and her husband a government worker, and in that instant I wondered how she could afford this luxury.

"So that's it? *Tangli Baiqi?*" Teacher Wang said, materializing before me at the trunk of her vehicle. "It's so large."

"Yes, yes. Yes, it is large," I said, gritting my teeth. I handed over the goods. "There's also the purse and two black *Maike Gaoshi* wallets, one orange and one pink, for your teacher friends."

Wang snatched the merchandise and thrust a wad of cash into my suddenly empty grasp. "Count it," she said.

I dutifully riffled through the stack, hoping no neighbor would spot me. The wad of bills—6,500 RMB in all—was as high as a stack of iPhones.

I finished counting. "Okay," I said, awkwardly.

Wang sauntered back to the driver's door and boarded her SUV.

I turned around, walked back to my apartment, and took a shower.

* * *

TWO DAYS LATER, my phone rang.

"The *Maike Gaoshi* bag is wrong," Teacher Wang growled in my ear. "The tag says USD$228—I gave you the equivalent of $298. Also, it's shiny leather and Teacher Li wanted the soft leather. We can't use it."

I was meeting with Amanda, and I marveled at the absurdity of the situation.

"Teacher Wang, I'll look up the receipt and call you later. I can't talk now," I said.

But Wang couldn't wait. "You took $298 from me. So you paid $298 for this bag?" she rang out.

"*Of course* I paid $298," I blurted. Did she think I'd tried to *profit* from this transaction?

"Then the store made a mistake," Wang said. "The store should send the right bag to China." Wang clicked off.

I imagined trying to describe this situation to Macy's return department. Instead, I explained to Amanda. "I'm shocked at how brusque she is," I said, though I recognized my complicity.

Amanda nodded. "Schoolteachers and officials in China are used to having parents do whatever they want," she said. "They have so much power." I glanced at my teenage friend, who was clearly embittered by her struggles with authority figures.

The following week I went to Pumpkin's school to fetch Mr. Kors. A guard approached. The man I nicknamed Smoking Guard usually dangled a lit cigarette from his lips, but today he walked briskly and purposefully, with not a vice in sight. Instead of swinging open the gate as he usually did, he stopped short and peered at me through the bars.

"You can't come over here whenever you want anymore," he whispered, a glimmer of warning in his eyes. "We have a principal onsite now." He retreated into the main building and returned a minute later, trailed by a thin young woman with a rhinestone buckle at her waist, clacking in four-inch heels.

"I'm the principal of this school," said the woman, looking at me through the bars, her gaze casting over me from forehead down to shoes as I shifted uncomfortably under her scrutiny. Finally, she nodded at Smoking Guard, who opened the gate. He avoided my eyes as I passed, and in that instant I realized he'd tipped off the principal, eager to get in with his new boss.

The principal's name was Kang. She escorted me into the same conference room where I'd surfed for Tory Burch the previous month with six teachers peeking over my shoulder. I glanced around as we stepped inside, half-expecting evidence of my wrongdoing to appear in its mocking white walls or reproachful, empty chairs.

Principal Kang was iron-fisted order, newly installed at Sinan Kindergarten, and it appeared she would put teachers back to work during school hours and relegate Prada and Michael Kors to nights and weekends. She ticked off her questions resolutely, the rhinestone buckle on her belt flashing in my eyes: How did I gain access to the school? What is my nationality? What does my husband do? What was my purpose of observing the classroom?

I answered her questions and told her I'm a writer who is researching the Chinese education system with a personal interest. "I have a child in the system," I told her. "My son attends Soong Qing Ling."

"I see," she said, and I detected a shimmer of awe.

We spoke more about my work, and finally she stood, ultimately unimpressed.

"If we don't call you, don't call us," she said, extending a hand to indicate the door.

"May I say hello to Teachers Wang and Li?" I asked, trying to carve my way back to the classrooms to retrieve Mr. Kors.

"I think you should leave now," Principal Kang said.

I felt Kang's eyes bore tiny holes into my back until I reached the front gates, and finally she disappeared when Smoking Guard swung open the gate. Only then did Teacher Li appear, her lined eyebrows raised, feet cycling in hot pursuit.

"Here it is," she said, chest heaving, pushing the Michael Kors reject into my hand. We both glanced over our shoulders in apprehension, but the principal had retreated into the depths of the building. Cheeks flushed, I extracted a three-inch wad of renminbi from my purse and counted out the equivalent of $298 in front of her. Smoking Guard watched me count, amused.

Li grabbed the stack. "The new principal does everything by the book," she whispered, and disappeared before I could respond.

Even Teachers Wang and Li had a master. As the iron gates clanged shut that day, the rejected purse in my hand, I knew that was the last time I'd see Teachers Wang and Li on campus.

A year later, when I met up with Wang at a noodle restaurant, she would reveal that the principal enforced rules no one had before. "We have to file reports, write papers. Stick to curriculum. I spend many more hours at school than before." Wang divulged another point of stress: Her twelve-year-old daughter was already preparing for the high school entrance exam. "She's studying until midnight every night already, and the test isn't for three more years," Wang said, glancing down into her noodle bowl, its steam rising, then dissipating against the worry lines on her forehead.

I asked how the girl was scoring in practice. "She's just average," Wang said. "An average student. She needs to score at 460 to get into the high school we want." Teacher Wang and her husband had taken jobs miles from their home in Yangpu District, so that their daughter could attend the highly ranked middle school near their cramped apartment rental. Only then, a full year later, did I appreciate Teacher Wang for the complexity of her situation.

She was not only an authoritarian propagating an unforgiving system, but also the victim of a new iron-fisted order. Above all, she was an anxious parent who was very much the system's subject.

* * *

BACK AT SOONG QING LING, Rainey's teacher awaited a response.

It had been weeks since Teacher Song's offer of "extra time" with Rainey, and I couldn't keep my head down at pickup forever. In the weeks since, I'd begun asking around, and I learned that evidence of her own corruption might have a visual representation.

"Just watch next time," a parent told me. "Watch her performance lineups." The order in which children are arranged for Chinese school performances—at holidays, for end-of-year song and dance shows—is supposed to be formulaic: Tall children in front, short kids along the back and sides, with particularly able or expert children slotted front and center. But Teacher Song's lineups were out of whack. Heights didn't match up, and children in front weren't always top performers.

In other words, Song was arranging spots by some other criterion altogether. Perhaps Teacher Song's participation on the gifting highway had a visual representation.

I didn't care much about Rainey's placement for school performances, although I knew little gifts and cash for services would certainly grease his way along his educational journey. Despite the upside, I knew that we simply couldn't participate. Rob felt the same way.

"I don't feel comfortable with 'extra help,'" he told me.

"A small gift of appreciation is okay," I agreed, thinking of pineapple cakes for the teacher, "but to pay our son's teacher for outside lessons is something else entirely."

"Let's do recorder practice ourselves," Rob concluded.

Rob and I sat together for a minute, in silent contemplation of our newfound commitment. Teacher Song was skilled, while Rob and I don't play the recorder. I was still recuperating from a childhood of forced piano lessons, and even though I was musically trained, I had trouble reading recorder sheet music, which looked to me like spherical beetles bopping across a paper. And Rob's adolescence had been spent playing air guitar to Led Zeppelin.

It would have been far easier to accept Teacher Song's time.

★ ★ ★

IT'S A STRETCH to say that every parent in China gives gifts, or even that most teachers and administrators accept them, but the practice is enough of an issue that the ministry announced that blanket prohibition on gifts, as well as the ban on exchanging money for a teacher's services outside the classroom. Many whispered that such anticorruption efforts only pushed the quid pro quo highways farther underground, where exchanges continued to thrive in secret.

What's clear is that many Chinese parents feel as conflicted as I do. One report found that nine out of ten parents would like to abolish Teachers' Day, which clogs roadways around schools each year as parents tote gifts to school. With government policy leading the way and parents' growing discomfort with gifting and gray areas, change was on the way.

Over the following months, though, I detected that Rainey was occasionally slighted by Teacher Song. Most instances were minor and unintentional, caused not by malice toward Rainey and more by her desire to favor another student. For the most part, I could live with it.

Other times, I had trouble controlling my instincts.

At one of the school's end-of-year shows, Rainey's classmate Li Fa Rong—whom a friend had begun calling "King of the Naughty Kids," since he was always being punished—misplaced his black performance shoes. Teacher Song's solution: Ask Rainey to remove his own shoes and offer them to the barefoot Rong. (The boys wore the same size.) I'd sussed out what happened after Rainey began to complain at home that his "new" shoes hurt his feet.

As I listened to my son recount this story, I became livid. The rage made my fingers twitch. I marched around our living room for ten minutes after Rainey went to bed, trying to breathe, and the next day I headed straight for Teacher Song at pickup.

I planted my feet before her. "I want Rainey to wear the shoes I bought for him," I said. Song was taken aback by my directness, but she continued, voice strong and steady.

"Rong Rong has no shoes. Can Rainey wear the extra ones I gave him?" Song said, indicating a pair several sizes too small for Rainey.

"They pinch his toes. And they aren't the shoes we bought," I said, chin lifted.

Song searched my face, deciding how to mount her rebuttal. "Are you sure about Rainey's size?" she said. "Rong Rong lost his, and we just want to make sure there hasn't been a mistake."

I pivoted, tramped home, and snapped a photo of the empty shoe box sitting in Rainey's bedroom, label clearly displayed. This was all the proof I had of his shoe size, since the footwear that had come in the box was being held hostage at school. I sent the snapshot over the WeChat group: "Rainey's shoe box. Size 33!" my message proclaimed.

To protect her *mianzi*, or face, before the group, Song could only relent. Rainey kept his shoes, and I rejoiced. When you can't find a work-around, you play what power you have.

Other times, I could only watch helplessly. Later that year, school officials in several major Chinese cities were caught sneaking anti-flu drugs into kindergartners' lunches without their parents' knowledge or consent. This was a money-saving measure aimed at keeping kids in school; should a student miss a day, the school loses the state-funded daily allotment for that kid. The opportunity proved too much to resist.

"Drug 'em up and keep 'em healthy!" proclaimed the headline of a news article about the incident. Hundreds of parents in Xi'an said their children had suffered headaches, body pains, and itching from being slipped the antiflu drug moroxydine hydrochloride. The same thing happened at a school in Jilin.

The central government immediately sprang to action; it launched a "blanket inspection" of all kindergartens through middle schools on the mainland, to suss out further corruption. Further *fubai*.

That area of inspection included Soong Qing Ling.

Rainey was sent home with a notice and a clear plastic vial the size of my thumb, inscribed with his name. The ten-milliliter vial in hand was made of thin plastic, and I pressed it absentmindedly. The plastic gave easily between my thumb and forefinger. A notice instructed us to send it back full of his urine for government spot checks.

I turned to Rob. "Wouldn't it be a huge deal if something like this happened in the United States? If school officials were drugging children?"

"Yes, this would be front-page news." Rob nodded.

"Legally actionable, right?" I asked.

"Legally actionable," he said.

"That's what I thought," I replied. Living in China sometimes does that to you. The truth can be so strange that it recalibrates the filter through which you see reality.

The next morning, Rainey and I marched to school with a vial full of yellow liquid. We never received word that any Soong Qing Ling kid had tested positive for antiflu drugs, and the scandal quickly passed from public consciousness.

10

BEATING THE SYSTEM VERSUS OPTING OUT

I no longer believed in Chinese morality. I'd been thinking in a way dictated to me by my parents and teachers, but it's not the only way. It's definitely not the right way.

—AMANDA

I got the phone call on a Tuesday in January.

"Is your son in the Middle Class grade?" said the voice, crackling with assuredness. It was the admissions office at the international kindergarten, where I'd submitted an application more than a year earlier.

"Yes, my son turned five this year," I responded, and suddenly found myself dancing a little jig.

"A spot has opened up for him," the voice said. "We will hold it for three days. If we don't hear from you, we'll move on to the next child."

Shanghai Victoria was tucked into a lane, down a peaceful, tree-lined street just a few blocks from home. There was no Blaring Bullhorn or growling guard. The principal met with parents regularly. Classes were taught in both Mandarin and English, which meant teachers were hired from both sides of the Pacific Ocean, and I found comforting the thought of a native-English-speaking, Western-educated teacher inside my son's classroom.

It was a popular school, a magnet for foreigners as well as elite Chinese who wanted something kinder and gentler than the local Chinese option. The school's marketing department knew this, and it was as smug as its waiting list was long. "We have an innovative learning and teaching style that helps children reach their full potential," the school's brochures trumpeted.

A Victoria parent I'd met was similarly self-assured. "Classrooms are *child*-led and the learning style is inquiry-based," she told me, drawing out the word "child" as if it were a strip of candy taffy that stretched and flattened without breaking in two.

"If a *child* has an interest in something, anything, they can broach it in class, and the class talks about it. *Child*-centered," she repeated.

"Not like what they do in Chinese school," she told me.

Victoria was as close to a Western-style education as we would get, and it was our escape hatch from Chinese education. For some reason, Rob and I hesitated.

"Overall, has Chinese school been good for Rainey?" I wondered out loud.

"I think so," Rob said, a pinched look suddenly crossing his face.

The Chinese way has been good in at least one aspect.

"When I grow up, you're going to be old," Rainey said, as he crawled into bed with us one morning. We three stared up at the ceiling, blinking the sleep from our eyes.

"That's right," I said, laughing.

"Who will take care of you?" Rainey asked, turning sideways to look at me.

"Funny you should ask, Rainey!" I responded, laughing.

"How about me?" my son said in a singsongy voice. He'd absorbed a little Chinese-style filial piety, and I liked imagining that he'd never deign to let me spend a day in a nursing home.

We had three days to figure out what we wanted to do. Three days was Shanghai Victoria's upper limit for indecisive parents, and I pondered

Rainey's educational future in a last, desperate effort to discern how he was doing at this juncture.

I still hadn't detected any worrisome behaviors, and I'd actually noticed a development I felt good about: critical thinking skills. Last month, Rainey had come to a frank realization about the Chinese teachers who control his environment: "Sometimes, my teachers tell lies."

I'd discovered Rainey's enlightenment during one of our bedtime pillow talks.

"Are you sometimes bad at work?" Rainey had asked me that evening, as we lay side by side in his bottom bunk, staring up at the wooden slats overhead.

"Well—sometimes I don't feel like doing something, and I don't do it right," I responded.

"What does your teacher say?" Rainey asked, looking small with his brown-haired head peeking out from underneath the covers.

"I have a boss. My boss will say, 'Do it over again and fix it.'" I answered. Rainey stared up at the bottom of the top bunk, and I continued gently. "What does your teacher do when you're bad?"

But Rainey didn't respond.

"I won't be mad at you, Rainey," I said. "Please talk to me."

My son's words came out tentatively at first. "Sometimes the teachers tell me I have to stay in the classroom, and everyone else goes downstairs to play," he said, gazing at the top bunk. "Sometimes they take me to look at the lower-class grades, and they say they'll put me here, and I'll never see my classmates again."

My pulse quickened. "Who else is with you when the teachers say that?" I asked.

"Last time, Li Fa Rong and Mei Mei and Wei Wei," he said, naming three other students. "They cry, but I don't cry." Rainey described their crying as thick, jolting sobs that rollicked throughout their bodies as if they were abandoned tiger cubs.

"Why do they cry?" I asked.

"Li Fa Rong believes that we'll never get to see our classmates again. But I know the teachers are faking it. I tell my classmates, 'It's *jiade*—fake!'" Rainey said, flatly, with little emotion.

"You know it's fake?" I exclaimed, glancing over at him.

"Yes, I know. I know they're lying," Rainey said, energized by my interest. "They also say I'll never see Mommy again, but I know they're lying, too." Rainey seemed particularly assured by this statement. After all, I am here, aren't I?

Most Chinese I know have a relationship with truth that's slightly different from my own. The Chinese are more likely to sacrifice the truth for a number of reasons, whether it's to pursue a goal, to protect the pride of an authority figure, to project modesty, or to preserve harmony. Studies comparing Chinese and Western culture back up this idea. Chinese doctors are less likely to tell a cancer patient the truth about a diagnosis, while Western doctors are more apt to reveal a grim prognosis. Chinese students look more positively upon a fellow classmate who told a small lie and more negatively upon one who told a truth that is boastful. A Chinese friend was told by her parents, "I found you in the garbage," to avoid having a frank talk about sex and procreation. In other words, the truth is not a worthy goal in and of itself, in part because the Chinese emphasize community over the needs of the individual.

In the case of Rainey's teachers, the women were bending the truth in pursuit of a goal that was good for the collective: children who behaved.

I knew somewhere, deep inside, that I should have been outraged, but I wasn't. I think I'd long ago realized the futility of anger about something that had happened in the past. Or, perhaps, I'd become inured to the extremes of the Chinese system, since my overriding feeling was one of celebration and relief that Rainey had recognized the threat as fake. He was questioning. Whether this was due to his home environment, or a quality of his personality, I didn't care. I simply wanted to celebrate: My son was thinking critically about his environment.

The tunnel of information between us was rarely this open, and I pressed as Rainey continued to stare up at the slats of the upper bunk. I gathered that Li Fa Rong—the "King of the Naughty Kids"—had been punished by isolation at least half a dozen times this year. Rainey had been sent away only twice, for infractions that included goofing around during lunchtime.

"Do you like school, Rainey?" I asked.

"I like school, but sometimes I'd rather be at home with you," he answered. Most days after pickup, he would run ahead of me, hand in hand with his classmates as they spilled out onto Big Green. He'd clamber over the playground equipment as I chatted with fellow parents and grandparents before we headed home. Rainey liked his friends, and I was also starting to find cohorts among the parents.

"That's normal, Rainey," I said. "I like work just fine, but most of the time I'd rather be home with you." As we lay there, a peculiar feeling moved into my chest, a swirling mix of uncertainty and guilt and love. It overwhelmed me, and a few tears emerged in the corners of my eyes.

"You know, Rainey," I started, "sometimes . . . many times, I don't agree with how the teachers do things."

"Okay," he said.

"But you have to remember—Mommy and Daddy love you for exactly who you are. We want you to feel like you can tell us anything."

"Okay," Rainey responded. We lay there for a beat. "Now what?" he asked, looking over at me.

"What do you mean?" I said.

"Should we go to sleep now?" Rainey said.

"Okay," I said, laughing, rubbing the tears from my eyes. "Let's go to sleep."

With that, my little boy turned to face the wall and his head relaxed into his pillow. I watched as he dozed off, his chest expanding with each intake of breath, and I found that my own breathing began to match the rhythm of his.

He'd been a comfort to the sobbing "King of the Naughty Kids," and he'd also become a comfort to me.

★ ★ ★

WHILE RAINEY WAS well into his journey of tiny adjustments, my aspiring Communist friend Darcy was showing mastery of the system.

"I had an interview at Jiaotong University," Darcy told me at our next meeting, naming one of China's top four colleges.

"An interview?" I asked Darcy. "I thought everything was decided by *gaokao*. By testing."

He grinned, gelled bangs bouncing as his head bobbed. "Some students have another way in."

The ministry was continuing to experiment with "quality education" and other reforms, and partly that meant chipping away at the importance of *gaokao* for certain groups of students. Make exams less important, the theory goes, and childhood becomes a little more palatable. Jiaotong University was one of about seventy colleges allowed to participate in "independent recruitment" that year, and the university had invited hundreds of students from all over China to "interview." Students who pass the interview may be awarded bonus *gaokao* points or be allowed to score at a lower threshold than normally required for admission.

I looked over at Darcy. "You were *invited* to interview?"

"Yes," Darcy said.

"Who chose you?"

"My school principal. I'm the only one representing my school. I'm *jijifenzi*—an enthusiast. They like me." He grinned, again. *Jijifenzi* literally meant "fanatic" and "opportunist" all rolled up into one. It wasn't necessarily a positive term, and in using it, Darcy was giving a nod to his own duplicity. But he was a Youth League member favored by school administrators. Darcy was well on the way toward accomplishing his goal, and in a Chinese student's life, the end often justifies the means.

"What did you talk about in the interview?" I asked him. He glanced down for a brief moment, shyly, before looking up again.

"I told them about you," he said. "I told them I was learning a lot about the Western way of thinking." I wasn't bothered. We each had our reasons for befriending the other.

Darcy sat before the Jiaotong professors, who had asked him to describe memorable moments and people in his lifetime. Darcy's carefully scripted answers revealed an accomplished young man who also had an understanding of humility and group welfare. It was textbook Communist humblebrag: Besides talking about his new foreign friend, he'd told the professors he revered the computer genius Alan Turing, but that "I'm too modest to have similar ambitions"; he'd gifted his mother a watch on Mother's Day, but it was purchased through barter, not cash; he'd rebutted his uncle—who'd lamented that China's standing in the world was *tai ruo*, or very weak—by insisting that China's "efforts to protect the country's interests and encourage growth are impressive, indeed."

I looked into the eyes of my *jijifenzi* friend. "Were you nervous?" I asked him.

"It wasn't difficult," Darcy said. "I answered some questions based on the truth, and others based on what I thought they wanted to hear."

This reminded me of his bone marrow story. Last year, administrators at Darcy's high school mandated that students should give blood for bone marrow typing when they turned eighteen. Administrators earned fame and recognition from government elders for creating a registry—at the time a novelty for a public school—but a handful of students rebelled. They were angry that their bodies were being used for public show.

Darcy thought the protests were silly. "For me, it's a must to be typed, because I will eventually join the Party. The school won't think I'm good if I don't participate." For Darcy, working inside the system would yield the best individual result.

So, he rolled up his sleeve for the nurse, and he didn't flinch when the needle slipped under his skin.

★ ★ ★

UNLIKE DARCY, AMANDA didn't care to rationalize anything.

She wanted out of the system. "I feel I have shackles on," Amanda told me. "I feel like I'm not allowed to think for myself."

"Can you be more specific?" I asked. Amanda thought for a moment before nodding.

"*The Merchant of Venice*," she said.

Amanda had studied Shakespeare's play in classrooms in both China and America. In it, the merchant Antonio borrows a large sum of money from the rich Jewish moneylender Shylock, seeking to help a friend in need. Unfortunately, the merchant's ships don't return from sea, and when the loan comes due, he doesn't have the money to pay. Shylock then demands a pound of flesh from the merchant per a contract clause, which is no simple matter: In sixteenth-century Venice, the surgeon's knife would take the merchant's life.

The action was straightforward. For Amanda, the drama came in the interpretations.

"Boy, were there differences," Amanda told me.

The Chinese verdict on the flesh-demanding Shylock is clear: The Jewish moneylender is the evil incarnate, greedy, and cruel. "The teacher always presented Shylock as a heartless and merciless capitalist," Amanda says.

Sitting in uniform inside her Shanghai classroom, Amanda never questioned this interpretation. She couldn't present evidence to the contrary even if she'd wanted to. In China, works of literature are presented only in snippets or even rewritten, then carefully selected for its message and pasted into a textbook.

Amanda's enlightenment had come in America, during her junior year abroad. There, she was assigned to read the full, unabridged version of the play. Immediately, she teased out the bigger picture: Shylock's peers ostracized and provoked him, a fact neatly excised from the Chinese textbooks.

"Shylock's behavior is repulsive, but everyone in society rejected him! Society created the 'hateful Jew,'" Amanda exclaimed, still astounded by this revelation a year after she'd stumbled upon it. "He's not the only party that is in the wrong." Inspired, Amanda promptly penned a high school thesis on how evil characters are partly shaped by the environment in which they live. Most classrooms in America will include at least some discussion of this sympathetic approach to Shylock.

"You could not have presented this viewpoint in school in China?" I asked her.

"No," Amanda said.

"Why not?"

"It deviates from the official interpretation."

"Well, then, how would you discuss themes in the play?" I asked.

"Discuss?" Amanda asked.

"Discuss, debate in class, talk about the different points of view," I said.

Amanda chuckled, a quick laugh that emanated from her belly. It was the first truly spontaneous expression I'd seen from her, and it was prompted by my naïveté. "There's no discussion," she said. "The teacher would just read an excerpt from the play line by line, tell us what each phrase means, and what we're supposed to memorize for tests."

"What about essays?" I asked. "Could you present your ideas in writing?"

Amanda shook her head. "There's no point in an alternative way of thinking. To get credit on any test, you have to write the 'official answer.'"

"What's the 'official' interpretation of Shylock?"

"That Shylock's evil comes from the lust for money, and lusting after money is bad," Amanda said. This made sense. The official Party line trumpets that money is acquired only for individual pursuits. The good Chinese puts country, community, and society before self, and the Party still held fast to that propaganda, despite the rampant capitalism and pursuit of money the Chinese witness all around them.

For Amanda, going to America had been like ripping the door off her birdcage. After she returned to Shanghai, armed with a penchant for caffeine and the indelible experience of a year in a US high school, she couldn't stop thinking about the "lightness" she'd experienced in the United States from being able to walk from math to English or from gym to world history at the bell. In China, students sit with the same classmates in a single room, all day long, for six years in primary, and three years each in middle and high school. Only the teacher at the front changes. America gave her a freedom to be different, and a certain liberty came from changing the faces around her every hour. There were also Kant and Schopenhauer, which she could read in all their unabridged glory. Amanda had tested a "couple years ahead" of her American classmates in math, so she could opt out of core classes that she found "easy," she told me. It was a statement free of arrogance; it was simply fact. She spent her newfound free hours immersed in Western philosophy: Nietzsche, Hume, Camus, Sartre.

"I no longer believed in Chinese morality," she told me. "I'd been thinking in a way dictated to me by my parents and teachers, but it's not the only way, and it's definitely not the right way."

The Chinese classroom began feeling like a "chamber of stagnation," with the walls closing in on her, inch by inch. Amanda began getting into trouble with her teachers.

Once back in Shanghai, Amanda was assigned to write an essay about the topic "The Unknown." She let her thoughts flow and her pen move freely, and out sprang the idea that people should respect what they don't know.

"I also wrote that people should challenge authority," Amanda told me. It was her alternative worldview making its grand entrance in her Shanghai classroom.

In China, the teacher crisscrossed her essay with red pen and scrawled a flunking grade across the top. A classmate adopted a system-approved template and received praise for "words that shine," as the teacher put it, which deepened Amanda's humiliation. "My classmate wrote beautiful sentences but empty sentiments, and she didn't challenge anything," Amanda told me. The Chinese judge good writing on emotional exaggerations and richness of language; by contrast, the American essay is valued more for stark language and clarity of argument.

The experience was painful, but formative. "I was really, really sad after that," Amanda said. The Chinese worldview no longer settled into her bones and instead sat on top of her skin like an ill-fitting uniform.

So, Amanda set her eyes on a new test. The SAT would lead her out of China and, if all went according to plan, into an American university. She'd had a taste, and she was all too keen to join a small but steady procession of Chinese marching out of the education system they'd been born into.

"As human beings," Amanda told me, "we have a psychological need to give our lives some meaning."

For every handful of Darcys rationalizing his place in the system, there was an Amanda scrabbling to get out.

* * *

EVERY CHILD IS born with an imagination.

It's this innate faculty that enables the process of creativity—which in turn gives rise to works of art, musical compositions, game-changing inventions, medical breakthroughs, new industries, or any novel or original approach that also has value.

As a toddler Rainey would dig for snails and root for stones, which he would scatter about in no discernible pattern as his giggles emanated across the lawn of our Shanghai complex. I reveled in watching him make discoveries. As Rainey's journey in Chinese school continued, the days of snail-digging and stone-throwing shriveled into a distant memory, and I grew more and more anxious that his environment might quash this innate quality of curiosity and exploration. Does Chinese school destroy creative and independent thinking abilities, or does it only temporarily discourage their expression? What are Chinese graduates like later in life, in high school and college, and after they enter the workforce?

May Lee teaches courses on innovation and entrepreneurship to Chinese university students in Shanghai, and she likes to try an exercise early on in a new class. The Chinese American instructor will invite ten students—all graduates of the Chinese school system—up to the front of the class. She gives each a paper bag and asks them to put the sacks over their heads.

"Then I ask them to take one minute to remove something they don't need," Lee says.

Blinded by a sack and standing before a hall of their peers, most Chinese students proceed to remove a watch, a shoe, or a coat. Others might take off their glasses underneath the bag, a piece of jewelry, or a sock. Meanwhile, their classmates in the audience are hooting with laughter, themselves unsure of anything but the humor in the spectacle.

After the minute is up, Lee will ask the students to remove the bags, and they do, standing at the head of the class, blinking under the lights of the room.

"I'll ask them, 'Why didn't you take the bag off your head?'" May Lee says. "And they'll look horrified, because it had never crossed their minds. There's an 'Aha' moment."

May Lee has conducted this exercise with thousands of Chinese students. By her estimate, only three of every hundred Chinese throw off the paper bag. Most students didn't have the mind-set to recognize, much less challenge, their basic assumption, she says: The teacher told them to put a bag on their heads, and therefore its removal was off-limits. May Lee didn't offer a control group, and the exercise itself could be considered something of a trick. Still, it got me thinking.

Would a teenage Rainey remove the bag from his head?

As concerned as I was about my own child, China's leadership was, too, about the students coming out of its schools. Creative and independent thinking skills are crucial to participating in the global economy, and many an economic expert says China risks stagnating as a second-rate world power if it cannot reform its schools and workplaces accordingly. The same series of global tests that gave Shanghai students top honors in math, reading, and science also found their performance slipped when it came to creative problem solving. "Could China ever produce a Steve Jobs?" is a lament that surfs the wave of many a Chinese conscience. "How much money have the foreigners made from us because they have better technology," a Chinese manufacturing executive bemoaned on national television in 2015, about China's failure to develop homegrown technology to produce ballpoint pen tips (triumphantly, a state-owned steel manufacturer declared in early 2017 it had precision-engineered the solution).

What did all this mean for my family? As I pored over research reports and talked to innovation experts, I discovered something sobering for Rainey's situation: The process of creativity cannot be separated from its social and cultural context, and, in fact, a person's environment is "more important than heredity in influencing creativity," as researcher Lorin K. Staats wrote. "A child's creativity can be either strongly encouraged or discouraged by early experiences at home and in school." I was also interested to learn that *practicing* being creative is critical to the process of creativity, ideally inside a setting that encourages its expression. Other academics purport that new experiences and a richness of stimuli are just as important to a child's mental development as good nutrition is to her physical growth, and that curiosity is a fragile thing.

Obstacles to the creative process litter the Chinese education landscape: Domineering teachers who discourage open questioning; exam metrics that keep children studying rather than exploring; social collectivism that promotes conformity; little time to practice being creative; and Confucianism's impact on the culture. In fact, the philosopher's very tenets discourage the "openness, opportunity, and cooperation that is necessary to continue to release creativity in China," writes Staats. A British professor who teaches law in China told me, "By the time the Chinese students get to me, they're not raising their hands. They don't even know how to frame a question. Their exploratory period has come to an end."

Furthermore, taking risks is punished inside Chinese education. My Shanghainese friend Ophelia learned her lesson very early on. Her first-grade math teacher asked the class to practice tracing equal signs with two perfectly parallel strokes. At six, Ophelia had trouble, so she tied two pencils together with string. Voilà! A single stroke yielded two perfect lines each time. Ophelia had taken a creative risk, but the teacher was livid. "She tacked my paper to the front wall and shamed me in front of the entire class," said Ophelia. An American friend of Rainey's had a similar experience. While tracing out a turtle in class, Myah swerved to draw a little flipper before looping from number 18 to number 19. Deviant appendages have no place in a Chinese kindergarten, and the teacher slashed up her worksheet with fat, red strokes. At home, Myah cried to her mother. "God wanted this turtle to have a foot," she said. Her mother thought for a moment and wisely replied, "God may want this turtle to have a foot, but in the school the teacher is God. So, in school you draw the turtle from number 18 to number 19."

The Chinese classroom encourages conformity and discourages experimentation, which can mean death for the creative process. The cultural fear of losing face, or *mianzi*, also rears its oppressive head: A Hangzhou high school principal surveyed his teenagers and found that most equated "thinking outside the box" with "not respecting teachers."

Such an environment can also percolate at home.

My parents certainly could have fashioned a more open childhood environment, as nurturing creativity and independence didn't much enter their thoughts. Independent children were disobedient children, and art, literature, or writing classes didn't figure into my parents' exacting calculus of how I'd get into an elite university. When we traveled, we rarely visited museums or art galleries, and European philosophy had no seat at the dinner table. For my parents—scientists with six college and graduate degrees between them—artistry was found in the curve of the sine symbol, melody in the memorization of the digits of pi, and free expression in the liquid nitrogen beads that scampered across the floor at my mother's chemistry laboratory. The liberal arts figured into my mother's life prominently as a hobby; she pored over English classics as well as science fiction and fantasy, but such things weren't considered something worthy of class time or career pursuit.

When it came to subjects that piqued my interest, my parents had their own ideas, rooted in the practicality of the job market.

"I want to study psychology," I announced during my sophomore year in college. I'd heard a fascinating lecture by the psychologist Anne Fernald, and I was enthralled by the study of mind and behavior.

"What kind of job are you going to get with a psychology degree?" my father replied.

"Well, how about sociology? Sociology is interesting, too," I responded.

"And where will sociology lead?" he rebutted.

A month later, I thought I'd found the perfect solution. "IE—industrial engineering!" I'd declared via telephone, naively triumphant. It was a field with enough study of human behavior to capture my interest, and plenty of math and hard science to placate my parents.

"Hmmmph—we chemical engineers joke that IE stands for imaginary engineering," my father replied. He'd fashioned exactly the type of environment to win that particular battle, and I spent many units at Stanford learning to design spillways for dams and calculate loads on truss bridges. I graduated with a bachelor's degree in civil and environmental engineering.

My Chinese and Chinese American friends who became sociologists, artists, and writers tell me that they had home environments that were nurturing and open, they said, which worked to negate the impact of school. (Or, like me, they found their path only later in life.)

Unfortunately, many students who end up at China's top-tier schools are skilled at following a well-defined pathway to success, which means many of the country's best first-year students haven't yet failed at anything (though this probably holds true at top universities worldwide).

In fact, some of China's highest-profile personalities insist they were successful because they had a third-rate—rather than a top-tier—Chinese education. Jack Ma, the billionaire founder of Alibaba, failed the *gaokao* three times and credits his inferior college education with his entrepreneurial success today. "If I'd gone to Tsinghua or Peking University I might be a researcher today," he'd said in a speech. "Because I went to Hangzhou Normal University, I got my cultural education by having fun. Kids who know how to have fun, are able to have fun, and want to have fun generally have bright futures."

Han Han is one of China's most popular bloggers. He's also a high school dropout. In a newspaper article that was reposted many times, Han Han likened Chinese education to "standing in a shower wearing a padded coat." "The very likely result of performing at full development (in Chinese education) is fully mediocre."

But is the Chinese education system all bleakness?

The creative spirit of the Chinese—while it may not always be on display inside the classroom—is buzzing inside China's marketplace, insist friends and colleagues who work in China. After years of oppression and decades of restrictive economic policies, the Chinese have been raring for opportunities to build a better life for themselves and their families, and even the tiniest bit of encouragement can unleash the floodgates of creative activity. The proof is a spirit that thrives in public chat rooms and inside start-up's, spurred by government policy that boosts venture capital funds and provides incentives to entrepreneurs. Out of the buzzing swamps of the Chinese tech world has emerged Tencent—which owns the WeChat messaging platform—and Alibaba, one of the most watched companies in the world. And when highly skilled Chinese migrate overseas, the results can be impressive: researchers studying entrepreneurship discovered that Chinese or Taiwanese immigrants founded

13 percent of Silicon Valley's start-ups and contributed to 17 percent of America's global patents, at a ratio disproportionate to their percentage in the general population.

I'm a product of American culture and, like my fellow Westerners, I romanticize the concept of a lone, eccentric genius in the model of Steve Wozniak or Steve Jobs, who invented the personal computer inside a garage or went on to disrupt entire industries. Who wouldn't want their kid to retreat with a couple of circuit boards and emerge with the foundations of a product that would revolutionize the way we live or work?

The Chinese way forward may be more interdependent and group-oriented at the moment, and change can be more incremental than earth-shattering all at once. Consider that Chinese painters are taught to model early work after a famous painter, and copy it over and over until the techniques are ingrained in the muscle memory. (Once the basic skills are mastered, true creation begins.) The Chinese companies considered most revolutionary today arguably took someone else's model, made incremental improvements, and eventually became uniquely Chinese. Is that not innovation, defined through a non-Western lens? For now this works. As a professor of entrepreneurship told me, if a Steve Jobs–like visionary is truly what China seeks, it needs only one. (In truth, there aren't many "Steves" in the West, either.)

A for-profit education executive told me that as much as Westerners—and many Chinese themselves—denigrate the authoritarianism in China's social, political, and education systems, it might actually prompt risk taking (though, it also clearly promotes breaking the law). "The Chinese are used to navigating brick walls—you build a wall, and they will have figured out how to get around it even before the cement is dry." A tendency to shun planning may also encourage the experimentation that gives rise to innovation; even cement in China is considered less permanent than it is in the West. Crews in China pour roads, only to have the cement chopped up two weeks later to install pipes. In this anecdote is a sense that anything is possible and the usual norms don't apply: Break, then pour again, and again, and again.

A strength of the Chinese marketplace is that change happens quickly, at a pace and fashion astounding to most Westerners who work here. In this environment, perhaps individual creativity, so valued in the West, is trumped by determination and opportunity. "If it's doggedness, the Chinese have that in spades, and willingness to try-fail-try-fail," said an American investor friend who's worked in China for fifteen years. Does this all mean that China can't produce a Steve Jobs . . . or does it simply mean the type of revolutionary change will just look a little different? A Western education professor who's been coming in and out of China since 1983 told me, "China's getting innovative and it's happening very, very fast."

For those who work in China, there's a consensus that this "no creativity" argument is a fallacy. The classroom certainly poses barriers to expression, among other problems (to be fair, educators around the world are also concerned about this problem in their own schools), but the Chinese are working to overcome these obstacles. If it's not happening in school, opportunities might come in the workplace. "Coaching and training are my top priorities," says a Chinese chief technology officer in Shanghai who has worked with thousands of young Chinese programmers and engineers. "It's my job to unwrap the chains. Some of them are extremely quick studies, and they become the cream of the crop."

Indeed, at its most effective, creativity and outside-the-box thinking require a solid foundation of technical skills and discipline, which holds true whether you're a painter, writer, or tech entrepreneur. When I think about Rainey's education in China, I know he'll come out with a strong foundation in math, science, and other academic disciplines. Ideally, he'll also have all those wonderful soft skills and an education in the humanities (which we may have to work on later). If the school environment emphasizes one side of this equation, as Chinese school often does, the other must be encouraged at home or outside the schoolroom.

I wanted my son to be conversant on all sides of the equation.

Rainey was making huge leaps in the areas that we liked about the Chinese system: He was a disciplined and polite child, and he was making progress with numbers, with more to come in primary school.

"Rainey's got a pretty amazing sense of self-control now," Rob said on the second day of Operation Decision Opt-Out.

"Yes, but have you ever seen him color?" I said. "He just can't seem to draw outside the lines, and he keeps drawing dinosaurs over and over again. Repeatedly." Though, some days I'd fret about Chinese education, and other times Rob would take a turn.

"That's not true at all," Rob said. "Rainey draws all kinds of things. And we're making sure he's getting opportunities to be creative at home."

"What about blind obedience?" I asked.

Rob laughed. "Does he seem like a weak-willed child?"

Just that morning Rainey had emerged at breakfast in an outfit he'd chosen himself. His brown corduroys clashed with his blue-and-white-striped shirt, and I was bothered by his sartorial choices.

"Morning, Rainey! Can you change your pants?" I asked, as I went to the kitchen to deposit my coffee mug. "You don't match."

"No," he responded, trailing me defiantly into the kitchen. "Did you hear what I said? I said, 'No.' I won't go change my pants." He planted himself in the doorway, holding my gaze.

Rob and I rejoiced in this display of will. "See?" Rob said. "And yesterday, he arranged all his dinosaurs in the form of an army to face off against his Star Wars characters. He called it 'Dinosaur Star Wars.' How do you make Star Wars even cooler than it is? Throw in dinosaurs. How creative is that?"

"You're grasping," I told my husband.

In truth, all seemed to be going fine. The longer a child stays in the Chinese system, I knew, the harder it might be to "undo" some of its more oppressive effects, and there might even be a point of no return for some Chinese children, but we simply weren't there yet. It's also possible Rainey is an outlier personality who might emerge from a Chinese environment with these qualities intact.

Not only had Rainey figured out adults sometimes lie—cue in Teacher Song's empty classroom threats—but by four years of age, he was already Santa sleuthing.

"Are reindeers afraid of Santa?" he'd asked earlier in the winter. We'd been decorating a pine tree from Oregon. The Chinese had begun to adopt Christmas wholesale, though they were more likely to celebrate commercialism than the birth of Christ.

"No, reindeer are not afraid of Santa," Rob responded.

Rainey thought for a long while, hanging a sleigh ornament on our tiny, ocean-container-shipped tree.

"But deers are afraid of people," Rainey challenged. "Does this mean Santa is not a person?"

★ ★ ★

AS I PONDERED all of this, Shanghai Victoria awaited our decision, and I considered some of the observations I'd collected along the way.

Darcy once told me China's education system was not yet strong, and only if the trunk grows well can the flowers blossom. The need to nurture and develop China's education system made the day-to-day experience of the individual student inflexible, he insisted. The British Chinese journalist Xinran gave me her own analogy: "China is very much like Picasso's paintings—we have eyes, a nose, lips, ears, but not in the right position."

Yet, with time and patience, those branches were beginning to stretch, and those eyes and ears were starting to find their place.

One day before I had to call the admissions officer at Shanghai Victoria, I embarked on one last gasp of an investigation. "What happens when a child misbehaves?" I asked a Shanghai Victoria teacher during a site visit I requested.

"I don't believe in punishment," said the English teacher I met with. "I just talk to them a little bit about what they did wrong." Punishment wasn't all that Victoria teachers disavowed. They rejected the traditional Chinese curriculum and instead embraced a free-form option where discussions could be designed around any subject a child broached. Free play was built into the daily schedule. Math was taught in a toddler-friendly fashion that shunned rote and embraced using numbers in everyday life. Management actively sought out parents to ask where the school might improve. This all sounded great, but were there any potential problems?

"Victoria's pretty good, but these parents concern themselves with things I don't care about," said my friend Alex, whose son attended the school. A trained attorney and a PhD with a talent for languages, she has a penchant for cultural affairs and worldly matters and zero tolerance for bullshit. I'd always appreciated her opinion.

"What do you mean?" I asked.

"These parents are entitled," Alex said. Victoria parents hailed mostly from America, Europe, and Australia, and to me, they exemplified Western parenting.

"At Rainey's school the parents work very hard to kowtow to the teacher," I told her.

"Not at Victoria," Alex warned me.

I once got a look at some of the emails circulating in a parent group at the school.

One mother of the Piglet class had a very distinct wish: "I wondered if we can request class teachers . . . give us simple live activity updates . . . a few pictures, video, et cetera." I'd always felt Western schools spent time catering to interests that have nothing to do with educating children. Not to mention, how could teachers teach if they were required to be daily documentary filmmakers?

But it was parental hubbub over the Victoria school's "shoe rack policy" that gave me the greatest pause. Chinese streets sometimes double as repositories for spit and cigarette butts, and the school had a fix: Children must switch their street shoes for a clean "indoor" pair each morning before entering campus. Before exiting at the end of the day, the children would again slip into their "outdoor" shoes, which had spent the day, germs and all, waiting on a shoe rack by the school entrance.

It was a worthy goal, but administrators soon called foul; teachers were spending too much time coaxing toddlers into and out of footwear. So the school changed the policy—without informing the parents.

Piglet class parents were horrified and incensed, and they mobilized by email: "I'm absolutely not happy," tapped one; the school did not "notify the parents officially," said another; and "The attitude of the school is quite patronizing in a way."

One grassroots Western mom took a poll to gauge support, receiving seven ayes. The mom intended to take this poll to the principal to enact change!

Perhaps I'd been in China too long, but I was appalled. Teacher Song would never feel compelled to notify Soong Qing Ling parents of anything she felt was strictly "need-to-know," much less *ask* permission if the school wanted to so much as remove a shoe rack by a door.

Once upon a time, I might have imagined mobilizing a parent group, say, to banish forced egg-eating or allow asthma inhalers to be kept near the classroom. I know now what would be in store for me. I'd have been exiled, if not physically, then socially. Other parents would whisper as I passed them on the Big Green. "That's the mom who tried to introduce democracy at Soong Qing Ling!" they'd say, avoiding eye contact.

Blaring Bullhorn—the principal who manned the gates with a speakerphone—would tell the security guards to shut the gates *before* I arrived.

The Chinese way is to hire good administrators and trust them to do their jobs; parents were to support the system, take responsibility for as much as possible, and keep petty distractions out of the equation.

I didn't disagree.

★ ★ ★

DECIDING TO STAY at Soong Qing Ling wasn't as much an intellectual decision as a gut feeling. I was growing more and more disenchanted with the system, especially with the political aspects of the classroom and the conundrum presented by gifting and favors. Yet the road out of Chinese education is strictly one-way—it would be impossible to backtrack after that first step out—and I didn't feel it was the right time to go. Many of the academic benefits granted the Chinese way would come in primary school, and any teacher or parent knows it's much easier to begin strictly and travel toward leniency than the other way around.

Thoughts swirling in my head, cheeks pressed against the iron bars, I stood outside the Soong Qing Ling gates to observe morning exercises a few hours before Shanghai Victoria's deadline.

A hundred children were out on Big Green.

"Line up!" Teacher Song called out to her charges, clapping twice, and her students filed into regiments alongside the four other classes of the Middle Grade. Morning exercises were usually a half hour of choreographed routines, performed in neat rows, which transformed Big Green into a field of tiny marching soldiers.

Today, each child held a pair of plastic rings, and as the music began, I saw that Rainey was no willing participant when the music began. When children extended their rings toward the sun, Rainey's shot down toward his feet. When some splayed left, and others flew right, Rainey's stretched skyward, lifted by the enthusiastic arms of an energetic little boy. When his classmates' bodies lurched toward the turf for the finale, Rainey's was a body in motion, hurtling across the courtyard to call on friends in Middle Class No. 3. He skipped in the narrow alley between the lines, zigging from one end to another, like a water bug skimming the surface of a pond. Perhaps the teachers were making a concession for an overactive foreigner, or maybe their leniency was a taste of White Bible's "kinder, gentler" approach filtering into the school. Whatever the reason, I was comforted by his display of individuality.

"Let's go!" Teacher Song barked as the music stopped, and the children filed into lines for the orderly walk toward the classroom. When the line began to move, Rainey leapt like a frog, hands jumping forward, feet following shortly thereafter. In this way he hopscotched into the building until his white ankle socks fell out of view.

"He was just running around, and jumping, and no one was paying attention?" Rob asked later when I recounted the story.

"That's right," I'd said. "I saw a little boy comfortable in his skin," I told Rob. We exchanged glances.

In that pregnant pause, Rob and I sped through our thoughts, histories, backgrounds, how we'd arrived at this place in time. Rob had grown up in a Minnesotan town of two thousand residents, with a lake that could be skated in winter and boated in summer, and teachers who were generally friendly and helpful. For his own children, he wanted exposure to a bit more culture, academic discipline, and rigor than he'd had growing up. I was recovering from a pressure-ridden suburban upbringing, but I'd never lost sight of the merits of academic rigor in the way that the Chinese delivered it. Rob and I comprised the perfect confluence of factors to choose a Chinese environment for our son.

"So we're staying," I said, giving voice to the conclusion we'd already reached independently.

"We've already done the hard behavioral stuff and he's survived," Rob said.

Why leave before the academics begins? In the next year, we knew that teachers would begin exposing the class to basic math concepts and Chinese character learning. I shared news of our decision with my mother-in-law JoAnn, a career American public school teacher, and she wrote, "After you make the decision just know you did the best you could—and look for the good in whichever school you choose."

What exactly was the "good" of Chinese education? I would suss this out for the next part of our journey.

Rob glanced at me. "You know—we have to be okay with the fact that sometimes we'll doubt our decision, that we'll feel like outsiders, and that we won't always like the way they do things."

I grasped onto this thread. "My bigger question is: Is he generally happy in class? Is he confident? I think he is." We would observe closely, and take it day by day, month by month, year by year. Rob looked at me intently, his eyes holding mine.

"And let's not be so anxious. Maybe we should relax more," Rob suggested, gently.

That directive, I knew, was for me.

PART III

CHINESE LESSONS

11

LET'S DO MATH!

Chinese children are superior to US children in every domain: number and operation, geometric shapes, problem solving, and reasoning.

—A CROSS-CULTURAL RESEARCH TEAM, ON THE MATH SKILLS OF FIRST-GRADERS IN NEW YORK AND BEIJING

Teacher Song has proposed the forbidden at Soong Qing Ling.

"We'll be teaching math through tests, in-class activities, and real-life application," she told a group of parents, assembled for the purpose of understanding her academic goals for the Middle Class year. Even though she'd uttered a violation of ministry reform policies against teaching academics to kindergartners, we fifteen parents huddled over a rosewood table, complicit, faces blushing with desire for toddler arithmetic. No one wanted to "lessen the burden" if it meant our kids would trail a faceless mass of competition, comprised of the roughly eighteen million babies born in China each year.

"The children will count numbers below twenty and learn about progress relationships, such as 'five is one step higher than four,'" Teacher Song told us, as we nodded eagerly.

Rainey was just a touch over five years old at the midway point of the Middle Class year. While I'd launched my frenzied quest to affirm our educational choices, he had quietly morphed into a proper student who happily sat with his Mandarin tutor each day, prepared his own backpack for school, and nodded at teachers as he sailed through the entrance gates.

After all the difficult cultural and behavioral adjustments we'd made, might we finally taste our reward? A number of upsides were keeping us inside the system, and with the help of educators and experts, I'd narrowed these down to a handful of features.

Math teaching was one of them.

Later that week, Rainey brought home a counting exercise, covered in the teacher's bright red slashes.

"Did you do this worksheet in class, Rainey?" I asked.

"Yes," he said looking forlorn.

On the sheet was a building with six floors, and each floor was subdivided into seven units. Each of the resulting forty-two squares was numbered in sequence—starting with 101 for the first floor, and 201 for the second floor—and blanks awaiting a schoolchild's pencil were scattered throughout the grid. This required a basic understanding of triple-digit numbers.

"First 101, 102, then comes what, Rainey?" I asked him, but my inquiry met with silence. Is this too advanced for a child so young? I wondered.

Clearly, Teacher Song didn't think so, and she scribbled a prescription for us: "Needs to work harder," she'd scrawled in red across the top of the page. She also broadcast a grade: six of eight possible points.

As I pondered this number, my son glanced up at me. Clearly, he felt I should shoulder some blame for those missing two points.

"Mom, why aren't you teaching me math and how to read and stuff?" he asked. "Long Long's mom is teaching him to read English. And Mei Mei's mom is teaching her how to add."

I'd begun working numbers into everyday conversations with

Rainey, but such forays usually fell flat, interrupted by inquiries about anything that happened to catch his fancy. *Rainey's attention span needs fixing*, as Teacher Song had told me.

My conversations with Rainey typically went something like this:

"Your baby brother is one year old, and you are four," I'd say. "When Landon is two years old, you'll be five. When Landon is three . . ."

"When I grow up to be Daddy's age, will I have a baby?" he'd interrupt.

"Yes, you can have a baby if you want," I'd tell him.

"Who will be my baby?" Rainey would say.

"You'll have to find someone to *have* a baby with."

"Baby Landon can't be my baby?" Rainey would say, pointing to his little brother wiggling nearby on the floor.

"No, because Landon won't always be a baby," I'd say, spotting an opportunity. "You will always be three years older than Landon. So when Landon is seven, you'll be what?"

"Will I need to marry someone to have a baby?"

I'd typically answer that, married or not, it's best to be in a solid relationship with someone willing to co-parent because raising a child—especially one who's always asking questions—is utterly exhausting.

Clearly I was no model parent when it came to teaching math. To make matters worse (for my ego), I came across a report comparing math abilities in Chinese and American kids as young as five. I devoured the study. What advantages would Rainey earn from his time in Chinese kindergarten? Would I be negating those with my laid-back style at home?

Lead researcher Jenny Zheng Zhou grew up in China and is now a professor in New York City. She and her team focused on students in the two international cities she knew best—New York and Beijing—tapping the children just a month into their first year of elementary school to minimize the influence of formal math classes.

Their findings were unequivocal: "Chinese children are superior to US children in every domain: number and operation, geometric shapes, problem solving and reasoning," Zhou and her researchers wrote.

As young as six years old, there's already a gap? I delved deeper. Zhou and her team had asked kids to perform a number of tasks. On the easy stuff, such as counting to ten and reading and writing numbers, American and Chinese scores were virtually indistinguishable. The gap became apparent with the more complex tasks. Chinese kids named many more 2-D and 3-D shapes. They scored two times better on addition and subtraction of numbers under ten. They completely trounced the American children on mathematical word problems, such as a question asking children to distribute objects equally among a given number of friends. Altogether, Chinese kids scored 84 to American kids' 60 points in mathematical knowledge.

Why was this? The researchers offered a theory. "Mathematics skill is present in all cultures," they wrote, "but it will develop to a greater extent in those cultures that value it more highly."

You'll see hints of this cultural value strolling through a park in China, said Liu Jian, a mathematician who helped develop China's national curriculum. "Just this weekend, I saw a grandma take her four-year-old grandchild to the square and collect small stones. They started counting together. In our culture, doing things like this is a part of childhood life. We've counted since we were young."

Darcy's parents taught him to memorize mathematical facts from a very young age. "It's very important for your son to memorize *chenfabiao*—times tables," Darcy told me. Head bent over a cup of coffee, he produced a pen and began outlining a nine-by-nine grid on a napkin. "Many children can do this before primary school. And in primary school math examinations—we'll have five minutes to do fifty questions."

As he started populating the grid, finger movements speeding up as he scribbled, I grew anxious. "It's okay," Darcy assured me. "Just make sure your son can memorize the basics. It will make the more complex concepts in school easier to learn."

Amanda had a simple explanation for Chinese math prowess.

"We're not afraid of it," Amanda told me. "And we start practicing really young." When Amanda was just three, her mother had started exposing her to math through puzzles:

> A wall is ten meters high. A snail climbs up five meters in a day, but when he sleeps he falls back two meters. How many days does it take to get up the wall?

While most Westerners are still toddling about in diapers, Amanda was already potty-trained. And she could chirp the answer: The snail reaches the top in three days.

My parents had always impressed upon me the importance of math, and my sister and I swallowed their mantra and carried it with us to Houston public schools. There, I always found it astonishing that large handfuls of my classmates were happy to trumpet the fact that they struggled in geometry or chemistry. Some even wore it as a badge of honor. In American high school culture, generally, coolness and proficiency at math or science were sometimes mutually exclusive; one particularly keen psychologist dubbed the social costs of academic success the "nerd penalty."

I'd clashed with my parents on many ideas, but we were aligned in our belief that math was critically important to learn. It's not that I was a geek who was cornered at recess or got picked last for dodge ball teams. In reality, I was a dancer with a triple turn, I spoke conversational French, and I became a captain of my drill team senior year, the ultimate crowning inside a Texas high school. But I also found satisfaction in working my way through a complex algebraic equation. I considered it comforting to know an answer existed, and I enjoyed the finality of landing upon it. I was also drawn to English literature and essay writing—which usually launched a journey of emotion and contemplation—but sometimes I simply wanted to solve for *x*.

Delving deeper, it turns out that early math skills are actually correlated with higher academic achievement later in life, as well as earning potential. Early academic skills, particularly in math, are "the single most important factor" in predicting later academic achievement, says a researcher who analyzed data from thirty-five thousand preschoolers. A London-based think tank found that "math skills developed during primary school continue to matter for earnings 20–30 years down the line." Math SAT scores predict higher earnings among adults, while verbal SAT scores show no such correlation (though I wondered how the study accounted for cases like me, who aced the math SAT but chose to become a writer, a profession in which paychecks might come once a decade).

There are also benefits unrelated to bank accounts; people with good math skills are more likely to "volunteer, see themselves as actors in, rather than as objects of political processes, and are even more likely to trust others," according to the OECD.

Math skills clearly matter for countries, and plenty of research findings read like gentle encouragements for America, as if it were a nation of bozos who haven't yet mastered long division. Boosting the math proficiency of American students to the level of Canada's or Korea's would "increase the annual U.S. growth rate by 0.9 percentage points and 1.3 percentage points, respectively," found a report at Harvard's John F. Kennedy School of Government. Automation and an increasingly competitive global marketplace would continue to pinch job opportunities for lower-skilled workers, trumpeted another. High-wage countries such as America must adjust, or risk growing income disparities and political instability, wrote the researchers.

Policymakers in China don't need to sound such ominous warnings: The Chinese are generally on board, and its modern leaders have long understood the link between technical skills of the population and the health of its economy (although the Soviet Union's 1957 launch of Sputnik did spur a reorganization in US science education). Deng Xiaoping called science and technology the key to China's modernization after the devastating Cultural Revolution of 1966–1976.

Jiang Zemin, himself an electrical engineer, promoted "rejuvenating the country through science and education" during his leadership years from 1989 through the early 2000s. Hu Jintao was a hydraulic engineer who pushed the idea of "indigenous innovation" and an innovation-oriented country over the past decade.

Today, China's ruling class is comprised mostly of scientist and engineers.

All this reinforced the choices I'd made for Rainey, though frankly, I could do without the Chinese sense of superiority on this topic. Zhou Nian Li, the early education professor I'd befriended, told me once about a visit to an American grocery store gone awry. She'd put $25 worth of goods in her basket, but the cash register had failed mid-checkout. This little speed bump utterly stumped the American cashier. "I told her I could tally the bill orally, but the cashier had to wait until the machine was fixed," Zhou told me, with a practiced ripple of her fingers that suggested she could recite the first thirty digits of *pi* while flipping somersaults. The head of assessments at a Shanghai primary school bragged to me that Barack Obama's law degree wouldn't pass muster in China. "In China, maybe Obama would be a mayor, but definitely not president," Ni sniffed, nose lifted toward the ceiling. (I didn't dare ask Teacher Ni for thoughts on President Trump.)

That got me thinking. Chinese families seemed to value math as a matter of culture—you might say Chinese society awards a "nerd bonus" rather than subtracts a "nerd penalty"—but what was happening inside the classroom? How differently did Chinese and American instructors approach math teaching? What would Rainey be learning in China as he got older, and what was the US equivalent? Was one method clearly superior?

To find out, I decided to conduct an observational study of my own.

In Shanghai, Teacher Ni invited me to visit the primary school where he worked; we'd been introduced by a mutual acquaintance. The following year, I would also visit schools in Minnesota, Massachusetts,

California, and New York, eventually deciding to focus on an elementary school in the Boston area.

The two math classrooms I chose to write about—one in the center of Shanghai and another in a high-performing school district in Massachusetts—would provide a window into two contrasting approaches. Of course, certain features about the culture and system mean a US-China comparison will never be apples to apples; for one, Chinese schools generally don't mainstream students with special needs, which keeps children of a single classroom moving at closer to the same pace. China also expects that kids will fall out of the system at every level, whereas America prides itself on educating every child. (The most sweeping recent US education standards are named "No Child Left Behind" and "Every Student Succeeds.")

Yet comparing is a useful exercise, James Stigler told me from his UCLA office one summer. Stigler is a psychologist who has spent his career studying education across cultures. "Meaning often emerges through contrast," he and his co-author Harold Stevenson had written in his book *The Learning Gap*. "We do not know what it means to work hard until we see how hard others work. We do not understand what children can accomplish until we have seen what other children the same age can do. So it is with cultures."

Pen and notepaper ready, I looked across cultures.

* * *

FIRST, CHINESE MATH.

Teacher Ni found me at the guard entrance of his Shanghai primary school midweek on a spring day. I followed three obedient steps behind him, down a long corridor on the top floor of the school's main building. In the school's massive courtyard below, I observed several hundred students taking their morning exercises, uniformed children mirroring the movements of their leader.

We reached the end of the corridor and stepped inside a classroom. "These are seven- and eight-year-olds," Ni said, brandishing a hand through the air. Tall and imposing, with a habit of dipping his chin to emphasize a point, Ni carried himself with the import of his position.

"Are these students academically advanced?" I asked, counting thirty-two heads. These children would be second-graders in America.

"No, they are normal kids," Ni responded, "average for Shanghai. Some parents taught them academics before they entered primary school, but now we encourage play when they're younger."

His words were "White Bible talk"—what I use to refer to any government attempt to make education kinder and gentler for its students—but I didn't believe for an instant that these kids had spent hours on the playground as very small tykes. The students sat on gray metal folding chairs, clustered in groups of six and eight around rectangular rosewood tables, chatting animatedly. A slab of clear glass anchored each, and I could see knees neatly tucked beneath, feet flat on the floor. All wore the red crossover necktie that is the hallmark of a Chinese schoolchild. The setup was classic for a traditional Chinese classroom: cold tile or cement floors and unadorned walls but for a framed, hanging red-and-yellow Chinese flag. For an American used to bright, carpeted schoolrooms, the Chinese equivalent often conjured up an empty icebox with a locked access door.

Teacher Ni and I took seats at the back of the long, narrow room, and an assistant deposited a Samsung touchpad in my lap. "You'll need this for class," she said. At precisely nine a.m., the master teacher strode into the room from an entrance near the front and the room immediately settled. A young girl at the second table instantly rose from her chair.

"One! Two! Three!" yelled the girl, her head nodding with each utterance. "Let's read the twelfth text *XiHuMingDi*. Standing by the West Lake, willow branches swaying . . . READY, *GO*!"

Seated and chanting in unison, the chorus of students echoed after her, quickly calling out the words to a popular Chinese poem. Then, without further ado:

"Class begins," announced Teacher Zhou, a slight woman who stood at the front.

"Stand up!" barked the lead girl. Thirty-one chairs squeaked as the students pushed back from the tables. I watched a tide of black-haired heads rise.

"Good morning, students!" said the teacher, facing her class.

The students responded, then pivoted to the back to find me and Teacher Ni.

"And good morning to all teachers!"

I'd been swiping at the tablet, and the spotlight was unexpected. I slid a little lower in my seat.

"Please sit," Teacher Zhou said, turning to the blackboard. "Yesterday we learned from a video about square numbers. If I ask you to demonstrate a number squared using dots, can you do that?"

"We can do that!" the students chanted, in unison.

"Begin!" the teacher pronounced. Thirty-two heads dropped over thirty-two electronic tablets, and I counted eighty seconds of silence. The teacher clapped sharply three times. "One! Two! Three!" she barked.

"Sit up straight!" the students chirped, adjusting their backs into ramrod position.

"Look up at the screen," Teacher Zhou commanded, as thirty-two heads pivoted toward the screen at the front of the classroom. Zhou pressed a button, and an array of black dots materialized on a grid up front. My Samsung followed suit.

"Student Number Two! Stand up!" Teacher Zhou commanded. At Soong Qing Ling also, I thought, students were often called by number and not name.

The boy rose to meet her challenge. "I've made a square number of four."

Here, Zhou began a rapid-fire stream of questioning, inquiries, and answers coming to and fro as quickly as a Ping-Pong ball bounding across a table. "What is this number?"

"Sixteen."

"How did you arrange it?" said the teacher.

"Four dots in one line."

"How many lines?"

"Four."

"Please sit." The student lowered himself into the chair, as if an invisible hand were pressing down on his shoulder. The teacher tapped on a screen, and another grid appeared. I marveled that a student's work would be subject to the judgment of thirty-one classmates at the tap of a teacher's finger.

"Number Twenty-Seven!" Teacher Zhou said. She was calling on students at random, placing classwork on display without warning.

A young girl rose from her chair. "I made nine. Every line has three dots, and there are three lines."

"Number Four!" Teacher Zhou said, another grid appearing, another child standing. The back and forth was quick, but I detected no tension or anxiety. What would happen if a student hadn't finished the assignment or marked a wrong answer? Finally, the screen flashed a paper with dots not fully formed.

"Someone is still working. Let's count together for him!" Zhou said as the student rose from his chair on cue.

In unison, his fellow classmates chanted: "One, two, three, four, five, six, seven, eight, nine, ten!" The boy stood and listened, intently. To me, his face showed no traces of embarrassment.

"How will a student catch up if he falls behind?" I whispered to Teacher Ni, at the back of the room.

"If he doesn't grasp the concept, the teacher will alert the parents and require that he *buyixia*—catch up, practice at home," Teacher Ni whispered back.

With her call-at-random approach to in-class exercises, along with scores on tests, the teacher could identify laggards and implement a plan for action. This was impressive, but in almost every other way, I found this classroom jarring to my sensibilities. Buildings in China south of the Yangtze have little or no heating, and this school was no exception. This room was tapered and long, which cemented the teacher as a distant authority figure, separated from her students by the interminability of space and the impersonality of desks. Students were called by number, not by name, which helped order the classroom but also reduced children to a collection of digits.

I surveyed the students in their tidy uniforms, heads vibrating with the effort of clapping. I glanced away for a moment and looked back.

Not a single student had imprinted in my memory.

* * *

HOW DOES THE American math classroom compare?

The following fall, a teacher in a Boston-area suburb invited me to sit in on a class. Massachusetts has always taken top honors on standardized testing; by this measure it's among the highest-performing states in America.

I stepped into Teacher Denise's classroom and was immediately struck by the warmth of the room. The walls were painted a bright aqua color, the floors covered with blue-gray carpet. It was the industrial variety more suited for insulation and sound absorption than for sinking toes into shag, but it lent a sense of comfort to the classroom. Bulletin boards with backgrounds of indigo blue and bright yellow were spaced along three walls, offering a graphical explanation of the Dewey decimal system, or tips on how to use one's imagination while reading a book. Where the Chinese classroom felt like a metal-and-glass shoe box, the American counterpart was an Easter egg basket, with grassy padding and an incubator to keep everything warm.

"This is a yardstick," said Teacher Denise, a calm brunette with a steady voice. Where the Chinese teacher was formidable, planting on two feet at the front of the class, Teacher Denise preferred to address her class with weight shifted to one leg, giving her a casual and approachable air. She brandished a length of wood before the class. "How many inches are in a yardstick?"

A few dozen nine-year-olds sat before Teacher Denise on the carpeted floor. About a year older than the Chinese students I'd observed in Shanghai, they were dressed casually in long-sleeved T-shirts, pullovers, and comfortable pants.

A boy in a sweatshirt shot up his hand.

"Matt?" said Teacher Denise.

"Thirty-six inches," said Matt.

"Good. Now, class, I have a meter stick here—see the difference? A meter stick is a little bit more than three inches longer than a yardstick. If you look closely at a meter stick, you can see tiny, tiny little lines," Teacher Denise said. "The bigger lines where the numbers are are centimeters, and the smaller lines are the millimeters."

So began a unit on the metric system.

"We have the US customary system, and the metric system," Denise said.

From my seat at the side of the room, the little group of students resembled a carpet picnic. Where the Chinese children sat in uniforms, pressed into rosewood tables, these kids expressed individuality in every way, down to the manner of sitting: cross-legged and upright, legs outstretched and extended, or leaning back casually, as if lounging on a pool chaise.

"Let's think about what I just said," Teacher Denise continued. "We have centimeters and millimeters. They both have prefixes. What other words that you know start with centi- or cent?"

"Cents!" offered a child.

"And how many cents make up a dollar?"

"One hundred!" said another.

"One hundred! Are we catching a theme here? What other words are made up of cent? DeShawn?"

"Centipede?"

"Century?"

"Yes! And take a wild guess—how many years in a century?" Teacher Denise asked, moving to sit on the floor, at eye level with her students.

"One hundred!"

The group moved through the same exercise with "milli" and "deci"—"Millipede . . . millennium . . . good!"

Suddenly, Teacher Denise rose. "Okay, we are going to get into our groups to work on measuring," she instructed, and the room erupted into a flurry of activity. I counted about sixty seconds before chairs and desks in the center of the room were transported to the edges. Teacher Denise produced a handful of meter sticks. A small group of students dispatched to monitors at the back of the room for FASTT Math drills—an individualized program of games that helps students with basic math facts.

For the next fifteen minutes, the rest of the students measured. They placed rulers against various objects: desks, chair heights, the length of a pen, a pair of scissors, the width of a textbook. In groups they set about their task, chatting animatedly, heads huddled together, as a designated leader jotted measurements onto a lined piece of paper. Slowly, their voices rose several registers, and Teacher Denise intervened to kill the small talk.

"Boys and girls, volume check! Hands in the air! Volume check!" Teacher Denise called out.

Each child lifted a hand in the air, quieting for just an instant.

* * *

IN THE CHINESE classroom, the students competed in a drill.

"Now let's *biyibi*—compare!" said Teacher Zhou, populating the screen up front with a stair-step diagram. The stairs began at the bottom left of my Samsung and escalated toward the right side of the screen. Each step cradled a number.

"This staircase is full of traps," Teacher Zhou said, planting herself at the head of the room. "Step only on the stairs that have square numbers, otherwise you will fall. For each square number you will get one star. Let's see who can get five stars! Begin!

The class bent over their tablets, and I launched into my time counting. I'd gotten used to measuring the seconds during such periods, marvels of quiet efficiency and productivity.

Twenty-five seconds passed. Teacher Zhou clapped sharply, and I startled in my seat, as I always did when serene calm is stamped out by a Chinese administrator.

"One, two, three!" Teacher Zhou yelped. "How many stars have you got?"

A chorus of students erupted. "Five! Five stars!"

"Now submit your scores." The students tapped on their screens, and a number materialized at the front of the room: 27.

"Ah! There are twenty-seven students who got five stars." That meant five children had lost their way, and Teacher Zhou scanned her tablet to identify a straggler. "Student Number Five? How many square numbers did you find?"

A boy stood, a lone figure in his red crossover necktie. "I have four stars—four square numbers."

"Let's do it together." Zhou summoned the group. "What have you found?"

The chorus of voices erupted: "16, 1, 81, 49, 100."

"Student Number Ten! What did you leave out?" A second boy stood for his public lesson, then promptly returned to his metal folding chair. I was astounded by the content of the exercise, and I whispered to Teacher Ni. "How come such young children have already started learning square roots?"

Ni laughed. "It's likely parents started to teach them when they were five, or they were sent to outside classes before they entered primary school."

"That's exactly what I'm afraid of," I exclaimed. "I'm not doing this with my child at home."

"Ennnh," he said. "Don't worry. We have a handful of students who are in a similar situation. I'll use a metaphor here: When others around you are dashing to the finish, the desire to run sometimes flames out." In other words, don't kill the desire to run by pushing too early. Again, these were White Bible words offered in the spirit of lightening academic pressure, although few parents would risk being the only ones listening.

Next up was another competition. "Let's see who can make more square numbers than the others," the teacher proclaimed. "Begin!"

I counted seconds again, and this time I kept going and going. Seven minutes passed before Teacher Zhou clapped.

"Are you making very big numbers?" Zhou said.

"Yes!" they responded, in unison.

"Are you excited?"

"Yes!" the students answered, voices overlapping neatly.

The teacher had complete command of the classroom. There was little room for inquisition, and I noted that at no time did a student have the opportunity to respond, as in, "No, teacher, I am not excited, and I don't understand."

"If I only provide you a sum of continuous odd numbers, can you tell me immediately whose square number is this sum?"

"Yes," answered the chorus of voices.

I sat in the back, gaping at the screen. In this ordinary Shanghai primary school, these seven- and eight-year-old children were studying:

If: $1+3+5+7 = 4^2$

$1+3+5+7+9+11 = x^2$

What is x?

"Student Number Twelve!" she asked, surveying her seven- and eight-year-old students. "Did you discover the rule?"

A boy rose. "It's the number of addends." I shrank a little lower in my seat, as if I might be up next if I inadvertently caught the teacher's eye.

"Yes," replied Teacher Wang. "It's the relationship between addends and square numbers! Let's count how many numbers we have here for the next one!"

A chorus of students ticked off: "One, two, three, four, five, six, seven, eight, nine!"

Teacher Zhou nodded, satisfied. Her white sweater billowed as a gust of wind suddenly blew through the classroom through the open door. "Through today's study, we found the beauty of mathematics," she said, nodding at her charges. On cue, they rose, chairs screeching as they scraped against the tiled floor.

"Today's class is over . . . dismissed," Teacher Zhou said curtly.

The students returned her gesture with a short bow. "Please have a rest, Teacher!"

"Please put away your pens and paper," the teacher said, as she retreated stage left.

I glanced at the children. They rose from their chairs, bodies slanted in the direction of her disappearing back, thirty-two heads tipped forward in the slightest nod.

* * *

IN THE MASSACHUSETTS classroom, the children had started in on fractions, divvied into groups of four and five. Teacher Denise crouched next to a girl in a cluster near the door.

"How many centimeters is five millimeters?" Teacher Denise asked, searching the girl's face. The girl was stumped. Teacher Denise prompted her again, gently.

"How about this—If you have fifty millimeters, that's how many centimeters?" Teacher Denise tapped on the worksheet in front of the girl.

No answer.

"Think about it like this—how many centimeters are in a meter?"

At this, the girl was confident. "One hundred," she responded.

"Okay, yes! So write that down. All right? All right. So now think about five centimeters—and how many centimeters there are in a meter. So what would be the fraction for five centimeters?"

The girl wrote a number down on her page.

"Okay, beautiful," said Teacher Denise. "So five of them would be what?"

Teacher and student huddled over the page, heads brushing against each other. For a moment, from a distance, they looked like friends.

"You have the idea, right?" Teacher Denise asked, glancing at the girl's face, the hope and encouragement clear in her face.

* * *

SPEND FIVE MINUTES in either classroom, and the differences leap out, a chasm in style and expectation, as well as a basic dissimilarity in what is valued: the group versus the individual.

The Chinese teacher was the center of gravity in her classroom. She expected full attention, and she got it. In a thirty-five-minute session she asked fifty-nine questions and called on half of her class at least once—by number, not name—and completely at random. This was her method of sussing out individual progress inside of a group-oriented class; she could easily identify the laggards. Much of her lesson was delivered in lecture format, with in-class exercises cemented by a timed competition. With lessons choreographed down to the minute, she covered a great deal of content.

The American teacher was far more approachable. She sat eye level with her students, called them by name. She rarely demanded their attention outright and deployed little tools to capture and hold their interest. She gave few orders and asked only three students to answer questions in front of the class; the rest were volunteers. In a fifty-minute class she jumped between lecture format, small group, and one-on-one interactions. While her children were working together in groups, Teacher Denise spent an entire eight minutes talking to a boy named Matt. She provided many opportunities for a student to say, "I don't understand. Please help," and she told me later that working with small groups allowed her to easily "see who's not getting it."

I also noted a difference in what researchers call "rigor, focus, and coherence," for which Asian math classrooms have always been praised. The Chinese teacher launched a lesson on math facts (square roots) and then directly led the students into a deeper conceptual understanding (relationship between square roots and addends). The American teacher's lesson was less focused on math than on measuring, and when a child failed to understand fractions ("how many centimeters is five millimeters?"), she reverted to asking questions the child already knew ("how many centimeters in a meter?") rather than pushing onward to deepen understanding.

The Chinese teacher commended no one, while Teacher Denise delivered praise throughout the lesson, including the words, "You're brilliant. You're so smart. Smart kids in here! Very good."

The Chinese classroom notched in with a teacher-student ratio of 1:32, and did not include students with disabilities. The Americans came in at 1:6, with one teacher and two aides for eighteen students (and included students with special needs).

Finally, I chuckled, more was expected of a Chinese kid's bladder: Anyone needing the bathroom had to wait until the end of class. The American kids were given more latitude with expression but less trust with urinary matters. A hall pass can be had at any time but must be signed out via roster.

"So if there's a problem with the bathroom, and we need to know who peed over the walls, we have a name," Teacher Denise told me, without irony.

Of course, a single class is only one part of a lesson that develops across days, and what works in one part of the world can't necessarily be transferred to another with the same effects. (Also, the Chinese teacher delivered a straight concept lesson, while the American one was working on a more "open math concept" such as measuring.) Even so, dropping into a moment in time, to me, revealed useful insights into education culture.

* * *

"FATHER PISA" CONFIRMED my observations about classroom dynamics.

Assigned this moniker by the Chinese press, Andreas Schleicher is the architect of the international standardized test PISA, which spawned hundreds of international headlines. He looked exactly the part: Tall and slender, with shock-white hair and a pepper-gray mustache, he was the poster image for global standardized testing. Schleicher was eloquent on conference stages and a rock star in education circles.

In 2015 and 2016, I traveled to Beijing to meet with him.

"Let's find a quiet place to talk," he said, shaking my hand as a throng of conference attendees weaved around us. Schleicher is a statistician by trade and German by birth, and over the last decade, to some controversy, he's cultivated the PISA into the gold standard for international education comparison. His pitch goes like this: "There's a lot we can learn from other countries' education systems, and PISA is the test that tells us which countries deserve a closer look."

When Shanghai students came in number one on math, reading, and science, the result challenged many Western beliefs: Smaller classes are good. Let children run around in open learning environments. Exploration promotes creativity. Eliminate poverty because it's bad for learning outcomes.

In essence, Shanghai turned much of this "wisdom" on its head, logging top scores while featuring exactly the opposite of the common American wisdoms, with such characteristics as large class sizes and authoritarian environments. Of course, critics came out and pooh-poohed PISA results: China must have cheated. Shanghai doesn't represent all of China, so how can we draw wider conclusions about the country's education system?

Schleicher had an answer for all of them.

"Sure, Shanghai isn't representative of China, but Shanghai today is China tomorrow," Schleicher said. "These people have spent a thousand years figuring out how to teach good mathematics. Don't you think there's anything we can learn?"

I played devil's advocate. "The Chinese spend an entire childhood taking tests," I said. "They're good at it. Might that help explain Shanghai's top-place finish?"

"Yeeaahhh," Schleicher said, and before long, I came to understand that this polite affirmation usually preceded a counterpoint. "That explains part of it. But there's so much more to why they do well. Teaching in math particularly deserves a close look because their approach centers on 'rigor, focus, and coherence.' The Chinese put a lot of cognitive demand on students, and very high expectations for every child. They teach a few things well, and they have a very good chance of advancing their understanding."

In essence, Schleicher explained, much of the time in a Shanghai classroom is devoted to deep conceptual understanding. "What's probability, what's space, what's the mathematical function, what's relationship?"

"What about the way Westerners teach math? Is there not more application?" I asked him.

"Yeeaaaah," Schleicher said, nodding. "In the US and many countries in Europe, many mathematics lessons are tied to little day-to-day problems: A mile-wide-and-an-inch-deep curriculum. So you use conceptually quite simple mathematics, embed them in a complicated real-world context, and then we think we are making math relevant to kids. But it is actually a very shallow representation of math."

This Western approach comes out of the idea that memorization and direct instruction are negative, Schleicher says, but in fact knowledge delivered this way can be a very useful tool. "The Chinese memorize what needs memorizing and use the rest of the time to go very deep in conceptual understanding. Then we are surprised that our (Western) students don't develop deep conceptual understanding that students in Shanghai do."

Yang Xiaowei arrived at the same conclusion. An education professor at East China Normal University, he recently visited eighteen schools in the United States and concluded that the American teaching approach is "good in theory, but doesn't work in reality." There's too much focus on making kids "interested" in math and on project-based and experiential learning, Yang told me. "Too little focus on directly teaching math."

The student-centered approach that Yang witnessed helps kids engage more meaningfully with subjects and better understand classroom content, its supporters say. Kids can also travel at their own pace. (Yet, while this type of instruction can be very effective, it also requires training and preparation to deploy it successfully.)

In fact, the "direct instruction" favored by the Chinese is better for early learning in many disciplines, especially those with "multistep procedures that students are unlikely to discover on their own, such as geometry, algebra, and computer programming," wrote two professors in a *Psychological Science* article. In early science education, especially, "many more children learned from direct instruction than from self-discovery." A 2015 OECD report found that teacher-directed instruction is actually associated with higher scores in science (and inquiry-based instruction with lower scores). Kids in countries with more teacher-directed instruction were also more likely to express interest in pursuing a science-related career. I most enjoyed the stark takedown of a group of education researchers who wrote in the *Educational Psychologist* that "the past half-century of empirical research" provides "overwhelming and unambiguous evidence" that minimally guided instruction is essentially a failure.

Another lesson I've learned is that the Western attitude toward math needs a little massaging. In my interviewing, I'd heard more than a few American teachers and friends say, "Oh, I was terrible at math, and I did okay in life." A Michigan-based education professor who had spent thirty years researching mathematics curriculum told me he's frustrated by the American attitude toward his life's work. "I see an attitude of 'mathematics literacy—that's okay we don't need it,'" he told me. "You'd never hear these same parents say to their children, 'It's okay if you don't learn how to read.'"

In Boston, Teacher Denise herself espoused a math-as-elective approach: "Students don't need physics or calculus on a regular basis, but we use math every day. As far as fourth-grade math goes, it's essential. Further than that? I think it's individual. It depends on what your goals are," she told me.

Few Chinese would ever dream of saying that.

The words of Jenny Zheng Zhou and her NYC-Beijing researchers rang in my head again: "Mathematics skills are present in all cultures, but it will develop to a greater extent in those cultures that value it more highly."

Certainly, the Chinese exhibit undeniable flaws when it comes to math teaching—toddlers should spend time on swings, not times tables—and the system overreaches with its level of difficulty in later years. A natural interest in other subjects, such as art, drama, literature, and foreign languages, shouldn't be overridden for the pursuit of high math scores. By many accounts, Chinese students struggle when asked to apply knowledge to an unfamiliar situation, or when questions deviate in some way from what's taught. But I admire the Chinese dedication to math in the primary years and the rigorous, teacher-centered ways in which they deliver their lessons. Excellence sometimes means a healthy dose of basic memorization and regular practice of deeper concepts.

As Schleicher told me, "You can never copy and paste an education system, but you can look at the features that make a system successful and see how you can configure them in your own context."

In Rainey's first year at a Chinese primary school, I would watch, slightly stunned, as he learned double-digit addition and subtraction at six years old. He began participating in timed drills, in which he'd have a minute to complete twenty double-digit addition problems:

$$56 + 27 - 32 = \square$$

$$74 + \square - 21 = 42$$

When Rainey first showed me this homework exercise, apprehension flooded my body. I envisioned my son becoming a nail-biter riddled with performance anxiety, and thoughts of Rainey as a robotic android who recited multiplication tables streamed through my mind.

The reality was different. Just as Rainey eventually conquered the gridded worksheet from kindergarten—103 precedes 104 and follows 102—he learned double-digit addition and subtraction. The concepts took some coaxing and cajoling at the start, but with the guidance of his first-grade teacher and assistance at home, Rainey succeeded.

Soon enough, he was helping to count out change at restaurants and asking after the price of sneakers, and I saw that he gained confidence from his success.

How much does a sandwich cost at the local café, if the intervals between prices are exactly the same?

Cucumber sandwich: 35元2角
Ham sandwich: □元 □角
Salmon sandwich: 36元8角

"Thirty-six yuan for a ham sandwich!" Rainey would chirp.

Perhaps children are more than capable of learning subjects like math, even from a young age. The first step is to believe that it's both important and possible for our kids to accomplish.

12

GENIUS MEANS STRUGGLE

Americans emphasize achievement without hard work. They believe in the concept of genius. This is a problem. The Chinese—they know hard work.

—XIAODONG LIN, PROFESSOR OF COGNITIVE STUDIES

In December, Teacher Song distributed an announcement via the class WeChat: "We have been rehearsing for weeks for the Soong Qing Ling Annual Show."

The Chinese love a good performance, especially one with costumes and photo ops, and the WeChat parent response was robust:

"Excellent, Teacher!"
"You are so brave and diligent!"
"I look forward to the day!"

I had a hard time faking enthusiasm for this one, as did Rainey. "All we do is practice, practice, practice," Rainey said, wrinkling his face into a scowl at breakfast.

Winter in Shanghai always dampened my holiday cheer. Self-quarantined in our high-rise on polluted days, entombed in smog, we missed our parents and siblings most intensely during this run-up to Thanksgiving and Christmas. Our family was alone in Shanghai for the holidays. All we had was the Soong Qing Ling Annual Show.

The performance would take place a week before Christmas.

A couple of weeks prior, I slipped into school to pay tuition and stumbled upon Teacher Song lording over an assortment of children in a second-floor coatroom.

"The reason I keep repeating my requirements is so that you can memorize them," Teacher Song commanded. "*Tingxiong!* Attention!" A dozen students, facing each other in two opposing lines, stood erect at her command. "All right? If we teachers work this hard, you need to work hard, too," Song barked, her voice sharp and high. "March with feet up high! *Yibei-qi*—Begin!"

I spotted Rainey. He'd been wiggling and jumping, but he heeded her command with the straightest spine I'd ever seen on my son.

The two lines of children marched toward each other, knees lifting toward the ceiling, staggered just so. When they merged, precisely one forearm length separated one child's passing shoulder from the next.

Two weeks later, Rob and I filed into the Soong Qing Ling assembly hall.

It was the kind of performance I'd often endured during Sunday Chinese school in Houston. Every so often, the principal there would decide it was time to "display" the children, so teachers would spend weeks planning performances, sports meets, and speech competitions for us. Apparently that wasn't enough offspring-gazing time for my own parents, who also ordered up ten years of weekly Chinese dance lessons, which, of course, tracked progress in an annual gala performance replete with silk gowns and fan dances. Generally, I had little enthusiasm for such parades aside from the friendships I made through them (though as I grew older, I cherished the memories and my parents' attempts to keep their culture alive in America).

As Rob and I filed into the hall at Rainey's school, my childhood flashbacks began. Whatever the locale or era, the Chinese school performance always brandished three things: a finely dressed adult onstage with a microphone, a hundred costumed children waiting in the wings, and gaggles of parents in the audience with cameras around necks.

Today, Rainey's twenty-four-year-old associate teacher Tao took the microphone, enveloped in a peacoat the color of imperial red, fake eyelashes fluttering, as an undulating ocean of parents amassed before her. The air was electrified with thoughts of dancing offspring, and Tao lifted her voice into an artificially high register. "The Annual Parent-Child Event . . ." Teacher Tao began, hips swinging with the effort, "WILL . . . NOW . . . BE-GIN! Welcome the children onstage with your applause!"

Instantly, a dozen parents sprung from their chairs and crawled with their cameras to stakeout positions in front, with bird's-eye views of the stage. The speakers blared music. Three children materialized from the wings and positioned themselves at the top of a rubber-mat runway, which stretched into the audience.

What unfolded was an expertly choreographed, Chinese toddler version of Milan Fashion Week. At the top of the stage, the trio of children struck a pose, middle child thrusting an arm victoriously into the air like a singer who'd just finished an opening act. Her backup Spice Girls snapped into position.

"Freeze! One . . . Two . . ." mouthed a teacher at the end of the runway, starting a countdown. Parents scurried to photograph the immobilized trio, chortling at their cuteness.

"Three . . . Four . . ." Teacher Puppeteer continued, as a bouncy tune blared over the loudspeakers, conjuring up images of syrupy-sweet, animated lollipops.

"Five . . . Six . . . now walk! WALK TOWARD ME!" Teacher Puppeteer signaled.

In this way, the show cycled through several dozen trios of children. From there, the teachers launched into a string of colorful song-and-dance numbers, which began to plod after the first few numbers.

Suddenly, a handful of children gripping ivory-colored plastic recorders assembled on a tiered bleacher. Rob sat up straight.

"Rainey can't play the recorder," he whispered, a rare moment of anxiety for my husband. I glanced at the stage. Our son seemed conspicuously absent, as if there were an empty spot on the bleachers that only Rob and I could see. A designated conductor child began swinging his arms stiffly to the tune "Mary Had a Little Lamb," and a cacophony of notes filled our ears. "He's probably the only kid who doesn't know how to play," Rob muttered, remorse creeping into his voice.

We'd passed up on Teacher Song's offer to help Rainey with the recorder, and our own, deeply dedicated effort to help our son in her stead lasted about three evenings. Meanwhile, Song sent by WeChat relentless, infuriating messages about the recorder: "Oh, the future will be more and more difficult if the family doesn't practice hard to keep up."

"We made a choice," I told Rob, my eyes on the collection of pert, exhaling children onstage. "Let's live with it." We survived that melodic reminder of our parental inferiority, and then watched as the show cycled through a reindeer dance and a Santa-and-sleigh musical number, followed by a colorful rendition of "We Are the World." My stomach began to grumble. Rob and I sat and watched, whispering to each other occasionally, but before long we began to stare blankly, our butts aching against the hard plastic of the chairs

No Rainey sighting.

Finally, Teacher Tao, in her swishing red peacoat, announced "The Xinjiang Dance!"

"Yes! This is it," I told Rob, confident in my prediction that Rainey would be chosen for anything strange or foreign.

"Yup. Rainey will show up in this one," Rob agreed.

Located in the far northwest, Xinjiang is commonly referred to as the spot on earth that's farthest from an ocean in any direction. It's home to high concentrations of ethnic minorities, particularly Uighurs,

Kazakhs, and Tajiks. China's state-run media prefer to characterize this area as lawless and remote, but in truth such groups bristle under increasingly restrictive central government policies, and the Party has cracked down on Islamic extremism and "separatist activities" there. News reports paint the region as a haven for terrorism and mayhem. A bomb on a bus? Xinjiang. A knife-and-bomb attack at a railway station? Xinjiang. A truck hijacking? Xinjiang. Xinjiang natives who leave and migrate to other places such as Shanghai are often looked down upon by the majority Han Chinese (who comprise about 90 percent of Chinese living in China).

Our Han Chinese teachers could have chosen our foreign son to dance as a reindeer or a jolly Santa. Instead, I spotted Rainey waiting in the wings, dressed in the vest of a Xinjiang minority.

"There he is! Our son the Uighur!" I whispered to Rob.

Someone had cut a curly mustache from black paper and taped it to Rainey's face, two dark, plump worms that danced on his upper lip. As the song "Raise Your Head Cover, Let Me See How Pretty You Are"—a nod to Muslim headscarves—blared over the loudspeakers, a gaggle of children filed onstage in tiptoes, right arms raised high in salute.

Soong Qing Ling teachers are nothing if not politically wise. There was no hint of ethnic strife, mayhem, or terrorism. Inside this school assembly, the people of Xinjiang were happy, pirouetting minorities, the boys in coal-black vests sewn with golden braids and sequins, the girls in taffeta skirts over white tights. A pounding, rollicking melody kept time, as the children launched a partnered dance, which conjured up nomads dancing around a fire outside their yurts.

The whole shebang was farcical, but I couldn't deny that the dance was a marvel in instruction, rehearsal, and presentation. Soong Qing Ling teachers had choreographed a routine down to five-second intervals, involving dozens of small children.

Weeks of thought had gone into choreography and practice, and the costumes were made to order.

* * *

TEACHER SONG'S DEBRIEFING unfolded as if China had just defeated Japan in the Olympic gold medal match in badminton.

"We spent two and a half weeks preparing for this performance," said Song, breathless, addressing a group of parents who'd trudged into a fourth-floor conference room as the children dispatched for lunch. "The teachers are very tired. It was very stressful and challenging to gather all one hundred twenty kids in the Middle Class grades. But you see how much progress our kids have made!"

Each practice method had a purpose, Song explained, designed toward a specific payoff: Rehearsing with other classes develops social skills. Rotating teachers requires children to adapt to new styles of instruction. Memorizing positions onstage enhances spatial thinking. "The quickest children can remember their positions after two or three times. They know who's standing next to them," Song said. "Others might need five or six times.

"But with effort, all of them got it!" Song proclaimed, with a flourish of her arm.

I thought about the math classes I had observed on the other side of Shanghai: Effort yields results. Here at Soong Qing Ling, those lessons were being imparted early, in kindergartner form. These children had learned to take instruction, rehearse week after week, and demonstrate their progress before a captive audience. The way Teacher Song described it, the Soong Qing Ling Annual Show had been a test, and the children's efforts had paid off beautifully.

* * *

DARCY WAS FACING DOWN his own effort of a lifetime. The National College Entrance Exam was only a few months away, and my Chinese high school friend had embarked on a study regimen more tightly plotted than a marathon runner's. He'd passed the "interviews" at Jiaotong

University, which gave him a points advantage on the *gaokao,* but he still needed to score above a certain threshold to be admitted.

"*Dandiao,*" he told me, when we met for coffee.

"What does that mean?" I asked. Taken individually, the characters meant "single" and "note," but they pieced together to form no word I recognized.

"My life is *dandiao,*" he said.

I reached into the recesses of my memory. Nothing. "I don't know that word," I repeated.

"If there is a picture with twelve colors," he says, searching for the right words, "and next to it there's a photo with only a single color, the single-colored photo would be described as *dandiao*. That's how I feel. Life is only one color."

Darcy was describing monotony.

His days at school were plotted with precision: Six o'clock rise for a seven o'clock school bell. Meals of rice, vegetables, and soup stolen in the time between classes, with fifteen minutes allotted for dinner. A six p.m. prep class held him for four hours, until a ten o'clock evening dismissal, after which he headed to his room for some shut-eye. He would start all over again within just a few hours. Weekends brought more classes.

"When do you sleep late?" I asked.

"Sunday. Sunday is my favorite day, my day of rest. But even then, I think about *gaokao*. This comes at the expense of everything else."

For good measure, his parents had ramped up their scare tactics, deploying the stories they'd used to frighten him into study since he was small. A favorite involved the true tale of an older cousin who'd tested poorly, and of course a litany of misfortunes ensued: no job, no girl to marry, parents living in poverty. The boy finally found work through a family connection, but he's seen as "inferior," Darcy told me once. "He didn't work for it."

What I found most interesting about the cousin story was that the language around the boy's failure was never, "He wasn't smart enough." Instead, his relatives clamored, in a legend that would surely echo down the generations, "He didn't work hard enough."

Darcy promised himself he'd forge a different path. *Gaokao* is a "platform for success—a milestone you must survive so people will recognize your ability to achieve," he told me.

"You sound wise," I told him.

"Study arms you with more opportunities," Darcy responded. "When people jump at the same starting speed, the higher you jump and the longer you stay in the air, the farther you get. Studying diligently provides me with a higher starting point."

⋆ ⋆ ⋆

THE CHINESE BELIEF system around effort is one of the most important things I've learned from my childhood and from my time in China. From the legendary hardiness of the Communists on the Long March—a bitterly difficult, yearlong series of Red Army treks covering roughly six thousand miles, which launched Mao Zedong's ascension to power—to the fortitude of the average Chinese student today, Chinese culture propagates the idea that anything worth accomplishing takes serious, sustained effort. This isn't to say the Chinese don't believe in luck or the fates as assigned by folk religion. Proper homage to ancestors can supposedly bring about a string of blessings, and of course another group of beliefs espouses that birth legacy can be determinant of fate: "A dragon gives birth to a dragon, a chicken to a chicken, and the son of a mouse can only dig a hole."

Yet overriding this is an intrinsic belief that anything is possible with hard work, with *chiku*, or "eating bitter." If there's a goal worth accomplishing, day-to-day life might be absolutely and miserably unpleasant for a spell. It's a concept that parents tell their children, teachers ingrain in their students, and China's leaders use to motivate their populace toward the goal of modernizing China. The concept reverberates in the classroom; studies show that for kids who score poorly,

Chinese teachers believe a lack of effort—rather than of smarts—is to blame. "There is little difference in the intelligence of my students," Teacher Mao, a Chinese language teacher at a Shanghai high school, told me, his voice unwavering in his conviction. "Hard work is the most important thing."

Conversely, Americans and Europeans are more likely to believe in innate ability. How many times have you heard a parent say, "I was never good at math, so how can I expect John to like it? He doesn't have the genes for it." There's a tendency in Western culture to believe, when it comes to academics, especially something technical such as math or science, that you either have it or you don't. "Asian and Asian American youth are harder workers because of the cultural beliefs that emphasize the strong connection between effort and achievement . . . white Americans tend to view cognitive abilities as qualities that are inborn," as a 2014 research study put it, starkly.

Anyone who studies psychology and education will tell you this is a dangerous mind-set. It's simply not true that a child's innate abilities explain gaps in achievement. Overemphasizing a belief in talent gives kids a free pass: "Why bother trying if I can't help it? I simply don't have what it takes."

This mind-set profoundly affects the way a person lives a life. Psychologist Carol Dweck has devoted her life's research to the debate between intelligence and effort, and she says effort wins. "The self-esteem movement said that making children feel smart and talented would help them succeed in school," Dweck told me, from her office at Stanford University. "But too much emphasis on 'smart,' and these kids aren't motivated. They don't persevere."

The winning approach is clear: Instead of telling a kid, "You're so smart," we should say, "Good job—you worked hard."

When I think about the Chinese emphasis on effort, I realize it derives in part from the modern experience. For decades, the nation has witnessed legions of youth study for the *gaokao*. This nationwide, herculean effort—in and of itself—attests that the Chinese believe a high score is more likely to be the result of sweat and labor rather than some kind of inborn intelligence.

This type of mind-set has special value when it comes to learning math and science. "You have to work hard to achieve," says Xiaodong Lin. "And the Chinese work hard."

Small, wiry, and full of energy, Lin is a Chinese émigré and a professor at Columbia University's Teachers College in New York City. In one of Lin's most widely cited studies, she divided teenagers trying to learn physics into three groups.

One group of students was introduced to some of the greatest scientists of our age—Galileo, Newton, and Einstein—and told of these men's very intense efforts to develop the theories that made them famous. Newton was presented as a man whose hard-grinding, everyday work ethic led him to develop his gravitational theory. Einstein developed the earth-shattering theory of relativity, but, the teenagers were told, he also tried for the last twenty-five years of his life to establish what's called a "unified field theory" of electromagnetic and gravitational phenomena. There he failed.

A second group of students was told only of the scientists' lifetime achievements and nothing of the process in getting there. The control group mainly learned about the physics they were studying.

The results were clear. For certain groups of kids, hearing about the tortuous struggle of the greats *increased* their confidence and interest in learning physics. It was a "Maybe I can do science, too" moment.

Those kids who heard about effortless genius were de-motivated.

"Those who believe that intelligence is a fixed entity give up or withdraw quickly when facing challenging tasks," Lin wrote. "People who believe that intelligence is malleable and can be increased incrementally with effort are more likely to hold learning goals in school."

Westerners could take a page from this playbook, Lin emphasized. "Americans think that if you have to work hard you're not a genius. You see it in the news headlines all the time: 'Extreme Success Without Struggle.'"

This attitude creates problems in the classroom, says James Stigler, the UCLA psychology professor.

"In America we try to sell this idea that learning is fun and easy, but real learning is actually very difficult," said Stigler. "It takes suffering and angst, and if you're not willing to go through that you're not going to learn deeply. The downside is that students often give up when something gets hard or when it's no longer fun."

The Chinese teacher who wants to present a *difficult* problem has it *easy*, Stigler told me. "The Chinese teacher can just say 'Work on it!' and the students will suffer, they'll struggle through it, they'll be uncomfortable. The Chinese have socialized their kids to put up with suffering and discomfort and all the things that are a really important part of learning."

During a Minnesota high school visit, I met a Chinese teacher who'd bumped up against a clash of culture while teaching Mandarin to American students. Short-haired, with an easy smile, Sheen Zhang was nothing like the image of an authoritarian Chinese teacher she had as a child growing up in Xi'an, the Chinese city famous for its terra-cotta warriors. I told her so, and she laughed.

"I started out very controlling, but I noticed that if I yelled, my American students rebelled," she said. "They talked back!" Sheen made other observations: Her students couldn't sit still for an eighty-six-minute class, parents complained when she assigned too much homework, and she had to "make the classroom fun and enjoyable."

Ironically, Americans are on the right track when it comes to athletics. "It's all about getting better, getting better, working harder," Jim Stigler said. "In sports, we're okay with competition and struggle."

And the American conscience is okay with rankings in sports. "A ninth-place finish simply indicates a runner should retool and continue training—a ninth-place finish doesn't reflect poorly on a person's self-esteem or worth," Stigler said. "But in academics, you don't want to embarrass someone by ranking them Number Thirty because 'It's not their fault.' In American academics, 'you either have it or you don't.'"

"That's too bad." At this, Stigler, sitting in his UCLA office, rapped his knuckles on his desk.

★ ★ ★

A BELIEF IN a hard work over talent isn't exclusively Chinese, of course, but it seems to be a philosophy more easily embraced by the culture. Another one is the concept of committing knowledge to memory—especially by methods such as rote learning.

Memorizing gets a profoundly bad rap in the Western world. It makes robots of children, the belief goes, or androids of students who can only recite upon command, devoid of any creative thought. This follows Western philosophy, which promotes the idea that humans are more developed than animals. "The mind is not a vessel that needs filling but wood that needs igniting," proclaimed a statement attributed to the Greek historian Plutarch.

Today's Internet-savvy world helps enable this approach, allowing us to go through life committing very little to memory. Why should we bother, when knowledge is available at the click of a button? If you want to recall Shakespeare's Sonnet 18, the capital of Ethiopia, or the first ten digits of pi, a search engine will immediately cough up the answer. We have facts at our fingertips, and, as a result, schoolchildren are doing less and less work committing facts to memory.

Here's where I go to the research, which declares this a dangerous trend. Real learning doesn't happen unless information is imprinted in long-term memory, the cognitive scientists say, and that transfer of knowledge into the storehouse of the brain can be accomplished in part through memorization and practice. Here's the key: Once a child locks away key information, he can free up the active memory for thinking deeply—and even being creative. British educator David Didau puts it this way: "It's worth memorizing certain things to the point that they're effortless, so then you don't have to think." American psychologist Daniel Willingham wrote that "the bigger storehouse of information a brain has, the better the brain will comprehend information coming in . . . thus allowing more thinking to occur." Expert problem solvers actually derive their skills on "huge amounts of information" and experience stored in long-term memory, one research team wrote in *Educational Psychology*.

In other words, you can't just look it up, Google it, or ask your neighbor.

In general, the Chinese way of education fully grasps this concept. Children are in a "golden period of memory expansion," as editors of a primary school textbook on Chinese classic writing put it. A 1998 Shanghai experiment found that primary school kids who spent twenty minutes a day memorizing classics could read more characters and had increased focus and concentration after a year. Memorization of important facts also teaches students discipline, and this approach goes hand in hand with Chinese language learning for the youngest schoolchildren.

It's not that memorization in isolation is the key; even Confucian scholars say the philosopher promoted active inquiry and thinking through inference. And it's important that bits of knowledge are interconnected and accessible: A recent phenomenon in education psychology calls for introducing "desirable difficulties" into the learning process, which helps a person retain knowledge for longer periods of time. The Chinese memorize the basics, cement a strong foundation of knowledge, and use the remaining time to progress to deep conceptual understanding. They don't learn multiplication tables through project-based learning, says the OECD's Andreas Schleicher, chuckling, "That's a waste of learning time. The opportunity costs are very high. We do many things in a way that's not very effective."

As a child, I spent many an hour memorizing times tables, the periodic table of elements, algebraic formulas, and script lines when rehearsing for plays. Every theater actor knows that once the lines of a play are embedded in the brain, the true emoting can begin. More than a decade after a trip to Slovenia, I still remember how to count to twenty in Slovenian—*Ena, dve, tri, štiri, pet, šest, sedem* . . . —since Rob and I had spent several hours memorizing and chanting (assisted by beer and chestnut schnapps).

In his own journey to memorize Chinese characters, Rainey is up to three hundred already.

"Look, Mommy," he said, pointing proudly to his stack of flashcards. We store them in an empty oatmeal container, and every weekend we take them out for a drill.

We sit there for ten minutes a day—okay, well, not *every* day—and look at them together. I flip a card, he recites. I flip another card, he recites.

大—big
小—small
山—mountain
甜—sweet
老师—teacher

In later school years, the Chinese commit to memory the first twenty elements of the periodic table, mathematical formulas and theorems, and historical facts, among others. Passages from classical poetry and famous writings are also important; my father can still recount the poems he learned as a primary-schooler. I once asked Amanda which ones were her favorite.

"*Jing Ye Si*," she said, without hesitation, or "Thought Upon a Quiet Night," by the Tang dynasty poet Li Bai.

"Can you recite it?" I asked her. Amanda immediately looked up, as if inspiration would drop from the sky. Then, very quietly, the words came tumbling out. Amanda spoke of bright moonbeams shining into a bedroom, while a little boy's head "lifts, gazing at the moon, and sinks back down with thoughts of home."

When she finished, I sat there for a moment, and the latte-sipping patrons around us dissolved into irrelevance. She paused for a moment, then spoke, her voice as soft as if she were ambling up a fragile beam of moonlight.

"I first learned the poem when I was a very, very young kid, probably primary school," Amanda said. "We memorized it in class. It's just beautiful, it has a rhyme at the end, and there's a moon in the sky. When I see myself look at the moon, I think about my hometown."

"What is the moral?" I asked.

"The poem teaches me how to deal with homesickness," Amanda said. "When I was in the US, I spent a lot of time thinking about this poem."

Ten years after first learning the poem, Amanda could still recite it by heart. From there, she could talk about its meaning and conjure up enough emotion that it soothed a pining for home.

Any Chinese schoolchild would be able to do the same.

On this, I fall short. I could recall only the titles of the few poems I'd learned in grade school, and I certainly couldn't recite any of them from the first word to the last. What's more, a few lines of verse didn't help resolve emotional struggles in my life.

In that instant, sitting across from Amanda as she lost herself in the luminescent moon of "*Jing Ye Si*," I thought: "What a pity that I can't."

* * *

I DECIDE ALSO to look into the Chinese approach to teaching. For the importance that the entire nation assigns to education, what are its leaders doing to prepare those who deliver it?

Plenty, as it turns out. The Chinese believe teaching is an art form that can be studied and improved, like the craft that it is. Educators are steeped in a tradition of videotaping classes, evaluating teaching methods, and asking colleagues to observe their own instruction and offer suggestions. "There's a sense you can actually analyze teaching, make judgments about its quality, and come up with ideas for how you improve it," said James Stigler.

In China, teacher training is built into the daily life of the school, and it is generally rigorous and regimented. The average new teacher in Shanghai might spend about fifty hours a month in professional development, on top of her regular teaching load, for the first three years of her career. From there, requirements gradually ramp down, though even the most senior teachers may still listen in on two classes a month and exchange ideas with peers afterwards. Teachers of the same subjects may be grouped to swap information, and those of different subjects may meet regularly to talk about teaching methods, as well as each student they share.

There's more: Individual schools and local and district education bureaus each have distinct training requirements, and the central ministry also has recommendations. Some include sending teachers overseas for training and cross-cultural exchanges, which keep ideas and curricula moving across borders. At East China Normal University—one of China's top training institutions for teachers—every student is encouraged to spend at least a year in a foreign country. Rainey's Teacher Song told me she has visited Australia several times.

And while choice is valued in a democracy like America's, China enjoys the efficiency of being able to send teachers where they're most needed. The most experienced teachers may be sent to the most challenging classrooms and seasoned principals to schools needing expertise, with incentives offered where necessary. Other programs pair higher-performing schools with lower-performers in mentoring relationships, sort of like giving a Big Brother or Sister to a school in need.

Most notably, to the benefit of everyone involved—especially the student—teachers specialize in subjects from the very first year in primary school. A first-grade math teacher teaches only math, while another might be in charge of only science. This means that kids are exposed from a very early age to instructors schooled at a high level of expertise and content knowledge.

Conversely, American public school teachers in the primary years are generalists; a third-grade teacher might oversee all subjects, including math, English language, and the arts. Teaching is largely private, with little expectation of collaborating with others or performing research to improve teaching practices. And, generally, rigorous teacher training tends to end—or continue only outside of school—once actual teaching jobs begin.

I liked the idea that Rainey's first-grade math instructor would teach only math, undergo rigorous professional development that is built into her school day, feel adequately supported, and work with her peers to continually improve her practice.

★ ★ ★

ALL THIS TIES into the idea that teachers are worthy of respect.

China affords teachers more status than any other country, a global education nonprofit found in a 2013 survey (though I had proof enough in the jitter that overtakes my hands when I talk to Rainey's teachers). In fact, teachers are equated with doctors in regard and earn similar salaries (although both professions in China are generally considered poorly paid and rely on "red envelope" gifts to supplement income). Roughly half of all Chinese would encourage their children to become a teacher (in spite of the poor pay). Less than a third of parents in many Western countries—including the United States, France, the UK, Spain, Germany, and the Netherlands—would do the same.

Deborah, an American teacher who'd spent two years teaching in rural China, told me that she'd never felt as appreciated, or as hopeful about the impact of her work, as the first moment she walked into a Chinese classroom. I had the same experience when I taught English at a Shanghai kindergarten two subway stops from home, taking on two classes a week for two years. On my first day, I was terrified, and I eavesdropped outside the classroom door as the head teacher prepped her charges.

"Eyes up toward the front. "Be *renzhen*—serious," she'd instructed her children. The kids were packed so tightly into rows that the backs of the chairs rammed into the knees of the children behind. They were quiet, faces forward. "After class I will call the names of the children I thought behaved well," the teacher said, surveying the class. Finally satisfied, she beckoned to me, and I stepped inside.

"Good afternoon!" I said in Mandarin, and the silence of the room was immediately dashed as twenty-eight children rose from their chairs.

"Good afternoon, Teacher!" twenty-eight voices chimed back, their force seemingly pinning me against the wall. In this environment, my lessons unfolded with perfect rhythm, like the inhale-and-exhale of an accordion belching scales, and by the end of the first month my students could count from one to twenty in English, chirp all the days of the week, sing the letters of the alphabet, and recite the entire text of *The Very Hungry Caterpillar.*

"Polar Bear, Polar Bear, What do you hear?" I'd ask the class, just before twenty-eight voices came screeching back at me: "I hear a lion roaring in my ear."

I'm not saying children need to be packed tightly into rows, facing front, and reciting while the teacher lectures, but it's certainly helpful when children are expected to listen to the teacher and pay her respect, with consequences for failing to do so. When educators don't abuse this respect, the setup can be very effective in making progress in the early classroom.

I'm also appreciating the habits that Rainey is developing. A year and a half down the road, as he started primary school, I would find that he prepares his own backpack for school. He sharpens six pencils himself, checks for an eraser and black marker, zips up his pencil case, and slots his English, Chinese, math, and reading comprehension books into his bag. When the teacher sends him home with a notice, he brings it directly to us for discussion. On those days, I appreciated that his Chinese teachers began instilling these habits and behaviors when he was a kindergartner.

One is simply showing up on time. I finally understand the purpose of Blaring Bullhorn at the entrance gates. Authoritarian, yes, but effective—once a child misses the gate and experiences being shut out of school, he is rarely late again. Punctuality is an incredibly important quality for a schoolchild; a 2012 OECD study found that truancy or tardiness accounts for the equivalent of losing "almost one full year of formal schooling" in mathematics scores. During elementary school, before our morning walk, Rainey would wait at the front door, urging along his laggard parents. "Let's go, let's go," he'd tell us. Attendance is also critical: American students have nearly double the hooky rate of the average OECD country, which is directly correlated to lower scores.

An emphasis on discipline in school is also carried home.

"Where's my desk, Mom?" Rainey asked me one day. I stared at my then six-year-old, dumbfounded.

"What do you mean?"

"You and Daddy have a desk, but I don't," he said, spouting an observation he'd learned from speaking with Chinese classmates. Most Chinese homes contain an area specifically designated for their children's study, and that holds true across Asia. More than 95 percent of fifth-graders in Taipei and Sendai, Japan, had desks in their homes, compared with 60 percent for kids in Minnesota, which happened to be the state chosen by the authors of the study. Chinese children aren't simply using a cleared-off dinner table or corner of a coffee table. "If there is a desk in an American home," the authors wrote, "it is more likely to belong to a parent than a child."

Chinese parents are also formally roped into their child's education, whether they like it or not. Generally, they must review a schoolchild's homework each and every day, and the same goes for tests. It's not enough to go through the motions, as they must prove their parental diligence; primary school teachers ask Mom and Dad to sign graded exam papers as well as booklets listing the day's homework, which the child then returns to school the next morning. It's a signed, traveling messenger of communication between teacher and parent.

Yet parental assistance has a time limit. Parents should be very focused on the primary school years, Rainey's teacher clarified for me, with the expectation that children can manage their own workload by middle school. "Habits are very important," Teacher Song told me, in an end-of-school-year meeting. "You start the children out right with parental guidance, and then their own *guanli*, management, takes over."

In this arena, at least, we'd succeeded. Rainey was on his way to developing habits of disciplined study that would last a lifetime.

* * *

OF COURSE, ANY one of these upsides of Chinese education can be taken too far.

It's only common sense that teenagers shouldn't spend hours a week committing facts to memory without a larger context. Teacher respect and authority are helpful, but not when that power is used to break children's spirits. A cultural emphasis on effort shouldn't mean that the work is more important than life.

On this last one, in truth, I've always needed a little therapy. My own parents were workaholics, and in the face of an amazing achievement, my sister and I rarely heard "Good work!" Instead, we were taught to look toward the next milestone: "What's next?"

I remember this most distinctly from the day I got my acceptance into Stanford University. Toward this milestone I'd toiled my entire seventeen-year life span, and one day our black iron mailbox spit out a plump, white envelope with a return address printed in cardinal red. It was my ticket out of high school, away from Texas, and straight out of my teenage years. Only, my father was grappling with the one-way fare and the freedom it would bring his firstborn.

"What makes you think you deserve to go to California?" my father screamed. "Do you think you're worth it?"

We'd been arguing about my future for nearly an hour; my father had envisioned an East Coast Ivy League, and a notch down on his list was Rice University, which had offered me a full merit scholarship. I'd placed myself under the epicenter of my parents' red oak dining table, and I felt a strange comfort with the grain of the wood running overhead. I could see my father's bare feet pacing back and forth past the head of the table.

"I'm going, I'm GOING!" I screamed, to the underside of the table.

"Stanford is expensive," my father muttered, almost to himself. Faced with the thought of his daughter's sudden departure, he was swimming in an ocean of fear and uncertainty, as well as the prospect of a tuition bill arriving in the mailbox every quarter for a school that wasn't his top choice. Chinese parents typically pay, but this also buys them a voice in the decision.

In the last few years of high school, our fights would build until they erupted—extended, volcanic blasts of anger and steam—followed by long periods of silence and avoidance. We battled over nearly every aspect of a teenager's life, from whether I could go on dates ("only for prom"), spend spring break at the beach ("not during high school"), try out for captain of the drill team ("if grades stay perfect"), or buy a new Dooney & Bourke purse every year like the rest of the girls in my circle ("waste of money"). I secretly dodged the rules I found most egregious, but our arguments about the rest grew so fierce that before long, we both began making small adjustments and concessions, lest our household erupt into outright warfare. I'd learned to sense where to press and when to retreat—but on the topic of where I'd spend the next four years, I wouldn't give up.

"I'm going! I'm going!" I repeated. There was something about Stanford—volleyball in front of the quad, palm trees lining the morning commute to class, a fierce community intellect that seemed to belie that California sunshine—that grabbed hold of my imagination.

"Can you decide just like that?" my father challenged me, his voice softening a bit.

"Yes, I can, and I'm going to Stanford," I yelled from underneath the dining table. "And there's nothing you can do about it."

We exchanged words for another heated twenty minutes, until finally both volcanoes quieted. As always, my father managed to get off the last word, tossed over his shoulder as he stormed out of the room. "You better be worth it. If I'm going to pay for Stanford, you better do something with it."

It's fair to say I didn't get a single pat on the back for all the late study nights, SAT prep, or hours labored over college admissions essays. My father was already focused on what I'd do with a degree I hadn't earned yet, a full six months before I would even step foot on campus as a freshman.

In retrospect, a little praise heaped on my childhood shoulders might have helped quiet some demons, the ones with sharp, haunting voices that many of my Chinese and Chinese American friends speak of. These little guys cling to your collarbone and whisper, as Amanda put it, "You're not good enough, you're not doing enough, someone's always doing more, and doing better."

With the benefit of age and hindsight, I know that not every moment of the day should be spent in pursuit of accomplishment (although I'm sure I could find at least a quarter billion Chinese who might disagree).

On the flip side, I know that I generally outwork most people, especially in the face of a task that seems impossible. Perhaps here, I've benefited because I never had a self-esteem angel assigned to my shoulder, whispering sweet words of encouragement. Success boiled down to a simple equation, and I could always summit the mountain with enough well-directed effort.

Most days, I'm thrilled we've given Rainey an opportunity to learn this lesson.

The day-to-day challenges of Chinese school weren't what we'd expected, but, surprisingly, through all of this, we have endured—you may even say we're thriving. Rainey embraces hard work, adjusts well to adverse situations, and has become an open and curious child. He has leadership skills, and he makes me and his friends laugh. This is a gift he'll carry into the future. He's a gritty, resilient kid, and he's thriving in the face of challenge.

Including the dental variety.

"Rainey has four cavities," said Dr. Ni Na, a woman of few words, except for the handful I didn't care to hear. "Two are so big he might need root canals."

"But he's only five years old," I stuttered, in a dental office that towered high above the streets of Shanghai. "Those are just baby teeth—they'll fall out anyway."

But they must be saved, Dr. Ni Na insisted, as spacers for permanent buds underneath.

Ayi was immediately mistrustful. "Are you sure it's not just the dentist wanting to make money?" she hollered, when I announced the news at home. A few months earlier she had explained the countryside abortion: "In China, if you don't want the baby, you go out and ride bikes, run, and swim," she'd said, pumping her arms and legs vigorously.

"Uh . . . I think the dentist is trustworthy," I said, hesitant to ask after countryside dental practices, especially after hearing about the rural Rx for unwanted pregnancies.

"Back home, we don't get baby teeth treated," Ayi told me. "They just rot away. It hurts and hurts until it falls out and then a new one grows in."

I told her that seemed like a cost-effective option.

Sedation and laughing gas aren't common in pediatric dentistry in China, and Rob and I were anxious about our son's first visit, so on the day of Rainey's next appointment, we both begged off work for a couple of hours.

"I love *zaojie*," our little boy whispered as he met his parents at the classroom door, using a school term that meant "early pickup." "I wish you could *zaojie* more." I noticed a lump bobbing up and down inside his cheek.

"What's that?"

"Teacher Liu put an egg in my mouth," he said. I peeked inside. It was a quail egg the size of a large marble.

"Does she do that a lot?" I asked.

"Sometimes."

"Is it always an egg?"

"No, sometimes it's a dumpling." Teacher Liu, a stout woman in her mid-fifties who always wore a white apron, was the classroom ayi who oversaw the kids' eating and sleeping. Last year, I might have found it odd that a teacher would line up children single file and insert food-stuffs inside cheeks.

Rainey had a countermeasure. "I'm going to spit it out," he said when we moved out of Teacher Liu's earshot, lump moving under his cheek. I notice my son is highly skilled at talking with an egg in his mouth.

"Do they let you spit it out at school?"

"No, but I'm with *you* now."

Good point. We crossed Big Green toward the school's front gates, and Rainey ran ahead to a cheery trash can, painted to look like a smiling mushroom. He carefully slotted his head into the mouth, looked down into the bin, and spat. The egg shot out and landed on some discarded tissue. He studied the pale orb, pulled his head back out, and smiled as Rob and I caught up with him.

"Ready! Let's go!"

The dental procedure was quick, and it was an immense display of resolve from a five-year-old.

"If the nerve is exposed, then we'll need to go in and clean out the nerve cavity," said Dr. Ni Na, dressed in a gown covered with scampering zoo animals. Rainey lay underneath a massive neck-to-ankle apron, lighted Star Wars sneakers peeking out at the end of the sheet. The tray next to his chair held a variety of metal instruments that were small and child-size, as if the usual array of dental objects had been subjected to an incredible shrinking machine.

Dr. Ni Na moved quickly and efficiently. "This will taste like a little strawberry! So you're going to be very still, right, Rainey?" The dentist smeared numbing paste on Rainey's gum line and inserted a needle right into his soft pink gums. She pressed firmly down on the syringe. In all, Rainey would need three shots of Novocain.

"If you feel uncomfortable, raise your left hand," Dr. Ni Na said. Rainey raised his right hand, feet twitching.

"That's your right. Raise your left," Dr. Ni Na said.

Rainey switched hands, all fingers wiggling maniacally in the air. His fingers were frantic, but his body was still as the second and third shots were delivered.

"Is that a needle?" Rainey asked, speaking past the instruments.

"It delivers medicine," says Dr. Ni Na.

"I hate it," Rainey said, cotton moving in his mouth.

"I know you hate it, but we kind of need it," said the dentist. It struck me that she'd just uttered a parable for the Chinese education system.

Hence began the parade of shiny, pointy things: Dr. Ni Na inserted all sorts of instruments in my little boy's mouth: a butterfly prop to keep his mouth open; long, tiny metal sticks used to clean out roots; a silver hook to maneuver chunks of cotton. One instrument was long and needle-thin, as if it could poke out the eyes of bumblebees from a distance.

Rainey's little hands were clenching and unclenching, but he didn't move. His little feet twitched and trembled, but he didn't move.

Forty minutes later, Dr. Ni Na removed a bloody cotton, and it was over.

"See you next time!" Rainey chirped, jumping down from the chair.

On the way home, Rob and I brimmed with hypotheses, slightly stunned.

"Is this because of Chinese school?" Rob wondered out loud.

"He listened well to an authority figure and he can handle pain," I offered. "He knows he can't always expect everything to be easy."

"Let's go back there again!" Rainey chimed in, Novocain clearly still in effect.

The next day, I pulled aside Teacher Liu at pickup, firm in my resolve.

"Please don't put any more food into his mouth at snack time," I said. "Rainey should eat only at lunchtime. Eating frequently is bad for his teeth."

"All right," she said, blinking at my directness. "What about cookies—no cookies in the morning?"

"*Especially* no cookies in the morning," I said.

The Chinese had figured out how to make their children expend mountain-moving amounts of effort, obey authority, memorize multiplication tables, and practice for weeks at bouncing a ball in competition with classmates.

But for oral health and the dangers of refined sugar, I liked my way.

13

THE MIDDLE GROUND

Maybe the hybrid of American and Chinese systems is perfect.

—LIU JIAN, A MATHEMATICIAN WORKING FOR THE MINISTRY OF EDUCATION

We always made sure to attend the birthday parties of Rainey's classmates.

My fellow Chinese parents expended mountains of effort planning these festive gatherings, and I loved that they opened a window onto changing Chinese parenting and cultural mores. (Each party was a mini-sociological study, served up with green tea and cake.) More and more, I saw that urban Chinese were embracing American and European customs: champagne-and-seafood brunches, caricaturists and clowns, European-citadel-inspired bouncy castles. Of course, Chinese touches were always present: A greeting line commanded by parent-hosts, a professional photographer, a karaoke solo by the birthday girl.

To me, these celebrations indicated a quickly shrinking, converging world. A decade ago, a Shanghainese girl might never have met an American on her home turf; today I am raising my children in China, and my son stands in her receiving line. At the same time, the Chinese are heading abroad to America, Europe, Australia, and other countries in greater numbers, fostering a global exchange of ideas more quickly than at any time in the past.

Sometimes, the best intentions are lost in translation.

One Soong Qing Ling mom prided herself on being well traveled, and for her little girl's birthday party she purchased a piñata. The only foreign parent there, I watched, enthralled, as a staple of my American childhood became a modern Chinese fascination.

Most American parents of my ilk know the piñata's party parameters: children filed into a single line a safe distance back; three swings per blindfolded batter; plastic bags at the ready for spilled loot. I witnessed none of that order here. A scrum of children thronged around the piñata—which took the form of Elsa, the main character of the Disney film *Frozen*—in an undulating mosh pit. The birthday girl whacked at Elsa's face erratically with a heavy baton, the bat whizzing far too close to noses and baby teeth. I sprang to action.

"Children, form one line and back away from the piñata!" I said, in Mandarin, inching into the throng with arm outstretched, seemingly the only parent who recognized the mortal danger posed by Elsa. But swings continued at random, baton passing from hand to hand in no particular order. "Let it go . . . let it go . . ." I imagined Elsa singing, before deciding on the opposite. I raised my voice.

"ONE LINE! ONE LINE! BACK AWAY from the piñata!" I urged, louder. This time I tried to position myself in front of Elsa. "And we should take turns with the bat."

Still, the children rushed past me like whitewater around a boulder, and as I was transported to the side, I felt a tug on my elbow. Rainey had popped out from the crowd.

"Mom. Mom! MOM! Don't worry about it, okay?" he said, in English, before the scrum swallowed him again. In this moment I realized Rainey had become Chinese in a way I never would be; for one, I lacked his easy comfort with enormous, surging crowds. I carefully retreated a safe distance back, and what ensued was China: pandemonium eventually finding its own pattern of order. Magically, no child took a blow to the face during fifteen minutes of swatting, and finally Elsa's arm fell off: The piñata tumbled to the floor.

I braced myself for the familiar rush of children scrambling for candy, only to find the kids wandering off, bored: Elsa was an empty, cardboard box, lying battered on the floor. For all her preparations, the birthday girl's mother had missed the point of the exercise: She didn't know to ram the piñata full of sweets and treats!

I wanted to chuckle, then to educate and inform the girl's mother, but here, Rainey again materialized before me, pulling my ear down to his level. By now, my son had far surpassed his mother's ability to skip seamlessly between one culture and another, and as a five-year-old, he keenly sensed when I needed guidance—or even an intervention.

"Shh . . . I know, Mom, I know. But don't say anything, okay?" he whispered, sparing me an embarrassing exchange. Rainey was skilled at saving face, and he'd nudged me to leave the mom in blissful ignorance.

Sometimes the best intentions need some massaging—or a little cultural translation (at an appropriate time).

As my reporting journey continued, I witnessed this phenomenon of adoption and interpretation happening inside Chinese education, too (with plenty getting lost in translation).

A few years ago, one of Beijing's top public schools decided it wanted its students to have "international eyes." I recognized features of the prototypical American high school in their efforts: Rankings would no longer be posted. Textbooks must be left at school. Class sizes were whittled down to twenty-five maximum. A mental health club was advertised as a release valve for school pressure. Students could choose electives, including swimming, rock climbing, and Frisbee. And the *zouban* reform, or literally, "walk class," meant students changed rooms for each subject.

Curious to peek behind the curtain, I rang the school to introduce myself, then flew up to Beijing.

"National Day School was founded in 1951," Teacher Dai Chong told me on a blazing hot spring day, greeting me at the guard gate. "The school's first graduates served as People's Liberation Army commanders." We passed a monument that conjured up a giant, machine-graded multiple-choice answer sheet: three diamond-shaped concrete blocks thirty feet tall, with a hole punched out of each.

He seemed proudest of the "walk class" reform. "Now our students can meet a batch of friends in one class, and meet different people when they go to the next class," said Dai, a plump man with a bowl cut who chatted animatedly. Amanda would have welcomed this change; she'd called her traditional Chinese classroom a "chamber of stagnation," as she sat with the same thirty-, forty- or fifty-odd students for an entire career of schooling.

Individuality and choice were also on the menu. "We have found traditional authoritarian Chinese teaching has many ill effects and deviates from the essence of education—which is to serve individuals. Our teachers are like friends—equals." I doubted authoritarianism could be legislated away, but I was heartened by the effort. For an hour, I jotted notes as we walked the grounds, Dai leading the way.

"This is the student union president. She was just elected," Dai said, stopping in front of a poster hanging in a hallway corridor, displaying a photo of a skinny girl wearing glasses.

"You have student unions?" I asked, surprised. I wondered whether these groups were arranged only for show.

"Yes," Dai responded firmly, when pressed. "Students can express opinions and make proposals, and the union communicates those to various departments." Students could also send complaints online, or sit with a headmaster at lunch hour.

"Has anything unexpected happened?" I asked.

"Yes," Dai admitted, squinting in the sunlight. "Students care about their rights. We didn't expect this. They complain to different departments and try to get their problems solved."

Teacher Dai seemed genuinely surprised, and I found his naïveté endearing. "Can you give me an example?" I asked. We approached the school museum, and Dai swung open the glass doors so we could take shelter from the heat.

"There was no hot water in the winter, and students were not satisfied," he said. We approached a large bust of the school's founder, displayed before a pictorial history of the school, which reached sixty-five years into the past and spanned three walls.

"Before there was only cold water?" I asked.

"Yes. We thought cold water could enhance the endurance and willpower of the students," Dai said, with a chuckle that, for the first time, hid a bit of nervousness. "But we're more humane now."

And the demands kept coming, Dai said. One group wanted recycling bins. Another targeted teacher perks, including elevators that have been off-limits to students, who have toted books up multiple flights of stairs for decades. "Now students and teachers ride the elevator together," Dai told me.

National Day students have even managed to lighten up *jun xun*, or military training, one of the most brutal physical tests that schoolchildren all over China endure. The students protested the drills as "inhumane and dogmatic," Dai told me, and the school nixed the most grueling ones.

"What are the limits to change?" I asked Dai.

"There have been some negative consequences," Dai said, slowly. "Students have begun openly breaking rules and disobeying teachers. China lacks a civil society, and when people are granted some freedoms, there isn't a compass guiding their behavior."

"Can the school handle these new freedoms?" I asked.

Dai Chong was silent for a few seconds before laughing, gently releasing the worry from his face.

"We will see," he said. "We will see."

★ ★ ★

CHANGE IS AFOOT at schools in major Chinese cities.

For decades, the education ministry has been patient, pumping out policies for reform like a talkative police sergeant that drones on and on while the crowd never listens. Yet change is happening at ground level, driven less by the sergeant and more by public schools that enjoy the funding, special designation, or political clout to experiment. Private schools also have certain freedoms (though Beijing is currently ramping up efforts to ensure they keep a government-sanctioned political curriculum). "China is in the process of reform, and we can see that there's nothing that should remain unchanged, especially institutionally," said Wang Feng, my education ministry source. "From our macro-policy perspective, everything needs to be modified."

As the grass seedlings of experimentation sprout, Chinese families aren't standing idly by. (Patience has never been much of a modern Chinese virtue.) Students are heading abroad for education in a steady pipeline that fattens each year, emptying out in the United States, the UK, Australia, and even Japan and France. Teachers and education officials are also going overseas for training, education, and work. These Chinese leave their imprint on their adopted communities and also return home with new ideas about education.

The pipeline runs *into* China, too. Nearly four million foreigners have come to study in China over the past two decades, hailing from more than two hundred countries. This shrinking-world dynamic is making for a global mix of educational ideas faster than anyone ever imagined, and this exchange is happening on every level: between individuals, communities, institutions, and governments.

This won't be easy; my own little family faced a reckoning as we grappled with new ideas and an unfamiliar environment, and so will the Chinese. China's is an education system steeped in thousands of years of tradition, yet confronting a period of unprecedented change. It's a school system whose methods are constantly challenged but whose results—at least in Shanghai—others wish to emulate. As Chinese educators work to develop areas they believe are deficient—such as creative thinking, leadership abilities, socio-emotional competencies, and other soft skills—they must confront some new realities: To succeed means a student might sometimes step out of line. In the process, test scores might fall, lines of authority be subject to questioning, and long-held institutions challenged. Will Chinese parents, teachers, administrators, and the government be willing to accept change, and what that may mean for society?

One educator summed up the problem for me: "The Chinese want creativity and expect their children to attend universities overseas, but they have problems with boys and girls holding hands," she said. Just as this American parent had hoped to pluck items off the menu of Chinese education—"yes" on math rigor, "no" for unblinking obedience—so do the Chinese when they ponder Western-style culture.

On the flip side, Americans and Europeans are feeling a little uncertain themselves, as politician after educator after politician lectures them about lifting student achievement.

"We need to take a strong look at ourselves in mathematics, particularly since we're beginning to see a downward trend across assessments," said Peggy Carr, the top official overseeing education statistics. "US students are running in place. We're losing ground," said John King in 2016, the US education secretary at the time. The chorus has become as loud as Blaring Bullhorn at the entrance gate, only fitting, as China holds a special place in this global dialogue. Shanghai students' top-place finishes were so startling, in part, because the test doesn't purport to measure the knowledge you've acquired but rather what kids can *do* with what they've learned. Few expected China's stereotypical rote learners to top the rankings, and it was a wake-up call for education watchers around the world. The British also bought into the clamor for action; various education officials set goals of a top-five PISA finish by 2020, and one policymaker flew in Shanghai teachers to school their British counterparts on how best to teach math.

Yet in America, at least, we tussle constantly over who can set standards, what they should be, and how best to hold accountable those who deliver education. This back-and-forth is stalling progress; it's clear where we need to head but the journey feels uncomfortable, as educators struggle with the perceived difficulty in raising academic levels—without sacrificing what the culture celebrates.

"How we differentiate from the rest of the world is this creative, out-of-the-box thinking, and if not stacking up on these tests means we need to 'teach to the test,' then I worry we will lose our competitive edge," said Jennifer Price, at the time the principal of the high-performing Newton North High School. "The bureaucrat's idea is they want to see scores, production of a child, the old assembly line. But how do you measure production of a child?" Minnesota social studies teacher Brian Steuter told me. Others worry that the individuality of the student will disappear, as progress ceases to have a human face, instead becoming a number measured by school and by state.

The schoolhouse grass, it appears, is always greener elsewhere.

I understand the concern, and I've experienced the same anxiety as a parent. As a journalist, I've searched high and low for evidence that creativity and critical thinking are quashed when we focus on our kids' academic skills, and I haven't found a direct link (though, clearly, they need time to experiment and explore). Yet I have stumbled upon plenty of research that suggests a strong academic foundation, couched in knowledge, enables higher-order thinking, and even the creative process. "You can't think about something you don't know—try it for a moment—and the more you know about a subject, the more sophisticated your thoughts become," said UK educator David Didau.

As we all wrestle with change, the world's education systems are gravitating toward convergence in what a "twenty-first-century" student should look like. It's a catchphrase that educators bandy about; we may not know exactly how to brew the magic potion, but we know what should result from drinking it. Technical abilities are important, but so are the soft skills such as leadership, creative thinking, and the ability to work with people you disagree with. "Certain competencies must be integrated into everything that's done, but saying that is simplistic," American education professor William Schmidt told me.

"How to do it is really, really complicated, and there's no data or real pattern of success."

"Maybe the hybrid of American and Chinese systems is perfect," concluded Liu Jian, the mathematician who studies curriculum for the Chinese education ministry. Xiaodong Lin, the Chinese professor at Columbia University, said that "the Chinese have gone too deep on the content, and the Americans are not doing enough. The systems have a lot to learn from each other." "The pendulum has always swung from East to West, and back again," as Chinese early education professor Zhou Nianli put it.

At the end of 2016, I attended an education conference in Beijing along with academics and government leaders from countries on six continents. I watched a Frenchman speak of teaching engineering to the Chinese, a Turkish reform director talk about obstacles to change in a Muslim country, and an Ontario education minister talk of infusing sustainability into her schools' curriculum. It was a whirling hot pot of dialogue, and after a heated debate about "innovation in education"—a zeitgeist phrase no one seemed to be able to clearly define—an education expert from Mexico finally threw up his hands onstage.

"The speed of change in the way we educate is staggering," the man said, before a ballroom full of attendees. "We will reach 2030 and none of the things we are talking about will be relevant at all."

At least we're having the conversation.

* * *

AS RAINEY APPROACHES his final year in Chinese kindergarten, soon to enter primary school, I am clear on my priorities. I want academic rigor, but I don't want him huddled over books every waking hour like his Chinese peers. I want him to learn to draw, play sports, and enjoy leisure, as well as cultivate a penchant for drama and literature and comedy. If he wants to hop, skip, or scramble on the way to retrieving a ball—diving over the couch and scrabbling under the kitchen table—he's welcome to do it.

I want that exact middle, that convergence, or what we believe the ideal twenty-first-century education should look like. Policy makers are inching the world's school systems toward an increasingly globalized future, but none have yet arrived in a satisfying way.

As we wait, I'm cobbling together my own solution. Few education experts the world over would dispute that in the elementary years, the content of Chinese education is robust and rigorous. The math curriculum is advanced and well developed, teachers specialize in subject areas from day one of first grade, and a schoolchild who stays in until fourth or fifth grade should acquire Chinese literacy and its nearly thirty-five hundred characters. I like the Chinese system's parameters for academic rigor. "If you have to focus scarce energy and resources, focus on the early years," as a Greek education expert told me.

We'll keep Rainey in the Chinese primary school as long as we can, all the while understanding there's a hard stop for his time in the Chinese system. I speak often with American, European, and Chinese friends in China lucky enough to have choices—whether by foreign passports, connections, or resources—and we generally agree the ideal upper limit in the Chinese system is sixth grade, possibly earlier, depending on the child. We'll pull Rainey out, especially as the unseemlier aspects—backbreaking levels of homework, a slow brainwash of political education, crushing pressure from entrance exams—filters in.

When the negatives outweigh the positives, we will alter course. Meanwhile we'll reap the benefits of a rigorous early education, while compensating for its imperfections as much as we can. My own childhood enmeshed the Chinese way at home with the American approach in US public schools; Rainey's upbringing does, too, pegged with the reverse influences for home and school.

Somewhere within these constructs, we hoped to find balance.

True, the Chinese have a fundamentally different way of looking at the world: When international education rankings are launched into the world, a group of critics always emerges with the accusation that such tests don't account for differences in culture or government, issues like inequality and poverty, or the presence of special-needs children in any particular classroom. This is only natural. Over the years, I've met a handful of global educators who explained away their students' poorer performance in some way. A Danish education expert told me that Denmark considered the Chinese math curriculum and decided to go another way. "Our education goals are different than theirs—we're looking to educate rebels," he told me, chuckling maniacally while socializing with stuffed suits at an education conference, looking every bit the rebel himself. A Russian education official told me PISA simply fails to measure what his country values in fifteen-year-old students. "We believe fundamental knowledge is more important than critical reasoning at that time, since eighty percent of our students eventually go on to college," he said, dismissing the test with a wave of his hand. The Americans also have their self-soothing explanations, and the "no Chinese creativity" argument seems to ring loudest.

Certainly, an American teacher wouldn't dream of shaming a student in front of his peers or locking a child in an empty classroom; indeed, she might be dragged into court. Meanwhile, the Chinese are bewildered by the American obsession with sports such as football, which keep preteen boys circuit-training on rubber tires during the hours their Chinese counterparts are drilling in algebra. Cultural differences are stark, as are disparities in what different countries value. (Not to mention that China's is a developing economy while Americans and Europeans have generally enjoyed prosperity and power in the last century.)

But that doesn't mean we have nothing to learn from each other. To their credit, despite their dismissals of the Chinese way, the chuckling Dane and the dismissive Russian had considered the Chinese curriculum, and travel to Asia regularly for education conferences.

And as we argue over whether such international rankings matter, China and other countries are working to equip their kids with the higher-order skills that are important for a rapidly changing world, and are also quietly demonstrating stellar math, reading, and science abilities, at least in major urban centers. American, Chinese, Indian, Australian, and French kids are now competing against one another for college admissions spots, and they'll later face off for jobs in a global marketplace for talent. Meanwhile, jobs for less-skilled graduates are moving to developing countries, and the opportunities that are staying in the West are being slowly replaced by automation, the experts say.

The Chinese will be catching up on all the soft skills that Westerners pride themselves on. It might take two, three, or four generations, but it's coming. Meanwhile, far more Chinese are learning English today—it's part of China's national curriculum—than the number of Westerners learning Mandarin, the most spoken language in the world.

I'm certainly not advocating increasing the competitive stakes in education by testing our kids the Chinese way. I'm saying we should consider doing things a different way. Try something that scares you—you just might be surprised. For me, that something started with shaking up my assumption about what's possible with my child and with his schooling. Chinese Lesson No. 1 was that kids are much more resilient than I ever imagined.

As we approached the end of the second school year, Rob and I continued to celebrate certain behaviors and compensate for others. In this way, we always had our finger on the scale. "Ultimately, I believe that family culture overrides school culture," says Corinne Hua, a Brit in Shanghai whose children studied in the local system through primary school. The former Beijing headmaster Kang Jian told me, "I'd estimate family's influence over character, morality, and affection of a kid at sixty to seventy percent. Family holds great leverage."

Recently, I'd been using my leverage to ensure Rainey didn't become a boy who asks for permission to do everything.

"Mom, can I decorate this?" Rainey asked on a lazy Saturday morning. We'd just built a clay model of a boat and set it afloat in the bathtub.

Rainey clearly needed some retraining. "It's your boat. You don't need to ask," I insisted, while my son shrugged.

Another time, his teachers insisted on holding hostage his favorite toy, which they'd requested each child bring to school. Rainey asked permission to bring home his velociraptor, only to be denied.

"Did you insist?" I asked Rainey.

"No, I didn't," he said, glancing up forlornly. "Teacher Song said no, so I just left it at school." I was infuriated at what seemed like purposeless authoritarianism. I marched into the classroom the next day at pickup, beelined for the toy box, teased out Rainey's velociraptor, and marched out the classroom door with my son's plastic animal under my T-shirt. (A pointless maneuver, since it's almost impossible to conceal a hard, pointy-clawed plastic animal, but the attempt to hide it made me feel less confrontational.)

On the way home, Rainey became anxious.

"Mom, did you ask the teacher for permission?" he said, right hand gripping his dinosaur.

"No. You know that I didn't," I challenged him.

"Did the teachers see you?"

"I don't think so. Do you think we'll get in trouble?" I asked.

"No—but you're supposed to ask," he told me.

"Not always," I responded, voice rising a register. "Not always. You're not always supposed to ask."

"Yes, you are, Mom."

"No, you're not. The teachers don't always know best, Rainey, okay?" I said. "But, um, keep that to yourself, all right?"

Rainey looked up at me, a smirk on his face.

Somewhere in there, Rainey was developing leadership skills. After enduring a gruesome round of entrance tests and interviews—including one that required me and Rob to stand before forty parent-competitors (and a principal with a bell) to deliver two-minute speeches about parenting, while our kids took math tests in another room—we chose a Chinese elementary school a block down the street. This school was known for its arts, and the academic pressure wasn't known to be so intense.

I'd watch as Rainey grew into his skin in first and second grades with a natural confidence and a social intelligence that no textbook could teach. He was also a nice kid, and eventually he'd be "elected" a class monitor by his peers. In Chinese class, he began writing characters on command and also quickly learned double- and triple-digit addition and subtraction. Soon, the class moved on to multiplication. Outside of school, we attended swimming, tennis, and soccer classes taught by American and Australian and Nigerian coaches. We also began frequenting museums in Shanghai, and reading English literature at home. (We also invited Chinese friends to birthday parties where the piñata was filled with candy.)

When he was six years old, Rainey would insist on being dropped off at a birthday party where he'd be the only child who couldn't speak French. We'd befriended the birthday boy—along with his Belgian father—through a local Chinese soccer league.

"I had a great time," Rainey told me afterward.

"Did you understand anything the other kids were saying?"

"No—but I still played Star Wars with them," he responded, grinning at the thought of light sabers and Jedis.

This incident made me proud. My son had parachuted willingly, alone, into an environment without the skills to communicate, and he'd had a fabulous time. More and more, I was having affirming moments like these.

⋆ ⋆ ⋆

AS RAINEY'S SECOND year in Chinese school drew to a close, I checked in with my Chinese friends before our annual summertime visit to the United States.

Darcy invited me to witness his day of reckoning as he sat for *gaokao*, the National College Entrance Exam, along with nine million students throughout China. I rose early on a Saturday in June to plant myself outside his assigned testing site for the millennia-old tradition of *songkao*—literally, "send test"—in which parents and grandparents usher their progeny off to the races.

As I approached the Shanghai middle school where he would be taking the exam, I spotted evidence of a society on hold: police cars blocking off sections of street, orange-and-white barriers arranged around the entrance, and throngs of parents and kids crowding ever closer, clutching spare pencils, tissue packs, and water bottles. Local shopkeepers emerged to bear witness to the spectacle. City buses came to a rest along this particular street, paying homage to this monumental day.

I stepped into the crowd as it pushed forward, a mass of bobbing black-haired heads belonging to everyone who might have a stake in this game—parents, grandparents, teachers, and students—and every so often a student would extricate himself, shoulders lurching forward, to step over the orange barricade and walk toward the school entrance.

I scanned the crowd for Darcy. The last time I'd seen him, about three weeks prior, he'd shown me a black fountain pen, handling it as gently as one might a blind baby mouse. He eased off the cap to reveal a shiny gold point.

"This is my special pen," he told me. I detected a blush, and I knew I'd stumbled upon a secret.

"What makes it special?" I said, fixating on him intently.

"Okay," he said, taking a breath. "A girl gave it to me."

"A girl-girlfriend? You have a girlfriend?" Darcy's parents prohibited dating, which—as many rules in China do—only served to force illicit behavior underground. The pair would arrange meetings by text on their banned phones and slip into each other's dormitories after study hall.

Brandishing the shiny pen, Darcy told me, "I'll use this to write my name, but I'll use my regular pen for the rest of the test." He held the instrument for a moment, a flicker of calm momentarily crossing his face.

At the test site, a procession of boys and girls grasped clear zipper pouches toting pencils and erasers, their ammunition for the upcoming marathon, which would take place over two days. As I watched them filter into the school, it occurred to me that some students enjoyed advantages a casual onlooker would never know. Darcy had passed his "interview" at Jiaotong University, so he'd need only score above the first-tier cutoff to snag a spot. That partly explained his calm; barring a catastrophic memory failure, Darcy was likely to succeed. His girlfriend, a history major whom teachers had passed over for this privilege, would have to "test in, straight up," Darcy told me.

Parents surged forward, a guard pushed back. "*Jiayou!* Add oil!" the crowd began cheering, as the children walked, growing smaller and finally disappearing into the school. "Don't be too anxious," "Go slow," "Check your answers," advised a group of teachers, who'd arrived to cheer on their students.

A guard approached me as I continued to scan the crowd for Darcy. "The Wheel of Fortune," he said, addressing my face of worry.

"Come again?" I asked.

"The Wheel of Fortune. It's out of your hands now. Don't worry, your child will test well," he reassured me. Did I look old enough to have a teenager?

At 8:38 a.m., Darcy popped out from the crowd. He wore an easy smile, and his dimples were out. I watched as he made his way toward the entrance. He'd been clearing his required score by a large margin in practice tests. The boy walking beside him showed no such calm; he stepped through the gates and suddenly pivoted in panic.

"I don't have water," he shouted in the general direction of his parents. A ripple of empathy passed through the crowd. It could have been anyone's child, and someone coughed up a spare bottle so he could proceed.

Gradually, the procession of students slowed, and a guard glanced at his watch. "Nine o'clock," he announced. "They've started."

The crowd milled about uncertainly for a few moments, then began to dissipate, disappearing into the shops lining the street or heading toward police barricades to rejoin the main road.

Parents, grandparents, and teachers, too, had reached the end of a long journey. All they could do now was wait.

Nineteen days later, at precisely eight o'clock in the evening, Darcy logged on to the computer as his parents sat in the other room, casually eavesdropping.

The cursor on the screen blinked a few times, then his score appeared: 474 out of 600 possible points. For students wanting to study physics, chemistry, and biology, the cutoff line for first-tier universities that year turned out to be 423. He'd cleared the hurdle by 51 points.

Ten minutes later, a text message arrived from his girlfriend, bearing a single number: 449.

"A little close for comfort," Darcy affirmed, "but it should be okay."

A week later, I checked in with Darcy to find his life anything but *dandiao*—monotonous. *Gaokao* was over. Dad had bought him an iPhone. He was learning to drive. A leisure trip with his best friend was on the horizon. Darcy had gotten a job bussing tables at a bar and lounge popular with expats, to learn about foreigners.

Here, an entirely new world was revealed to him. After his first shift, he regaled me with stories of Americans who dropped $15 on a single cocktail; Germans who painted their faces red, yellow, and black (it was World Cup season); and seating so freely arranged that the oldest and most important person wasn't required to sit at the head. "It was so lively," he said with a smile. "In America, outgoing people must be more attractive."

With this knowledge, he'd embarked on a plan to change his personality. "I want to be more gregarious," he told me.

His father soon stepped in to foil his plans, demanding he quit bussing tables on day two. Darcy's face was full of disappointment. "My father told me I needed to save my strength for college," he told me.

His official acceptance letter had come from Jiaotong University. "A journey of a thousand miles begins with a single step," it began. Darcy had just completed a major educational milestone, but already he was looking ahead to the next mountain.

★ ★ ★

AMANDA GOT INTO one of America's most prestigious universities, celebrated for its Nobel laureates and Fields Medal mathematical wizards.

I expected momentous celebration, but instead I found a depressed and uncertain teenager. Amanda's mother was in the middle of treatments for Stage III cancer; she'd been diagnosed when Amanda was studying abroad during her junior year. The family buried the news until her return to China almost a year later: Truth always took a backseat to education.

Yet it wasn't her mother's illness that got Amanda down; it was the state of her relationship with her parents.

Amanda broke down on our next visit. "I feel like all I've done in the past ten years is to please my parents or pay them back," she said, wiping away tears. "I used to be part of my mother, but now I should be an individual."

Her tears were those of sudden realization: She understood her mother and father were never going to let her go. The umbilical cord she thought would be severed when she set off for college would now stretch from Shanghai all the way to the United States. Amanda's parents wanted daily text updates and phone calls, with a heavy hand in choosing her major and her career path forward. I thought about the fights I'd had with my father just before I was to leave for college: Was my life his to direct—or mine to own?

"But you'll be out of their reach next year!" I told Amanda, trying to sound cheery. "Will you really text them every day? You don't have to . . . you don't have to at all!" After heading to Stanford at eighteen, I fielded a phone call from my father nearly every other day, huddled over a phone that I pulled into the hallway at odd hours. I let my father choose my field of study, my summer internships, and even my first job, until finally I put up my hand. We didn't speak for a year, then abruptly, as if he'd suddenly realized that having me in his life meant loosening his grip, he let me go. (Guide the way, but don't suffocate, he later confessed to me after I became a parent.)

"What choice do I have?" Amanda responded, her voice low and quiet. "Their happiness depends on my happiness."

That year, Amanda had begun reading *Escape from Freedom*. In the writings of the German philosopher Erich Fromm, Amanda found parallels in his ideas about a person released from a power or convention. "Most individuals harbor the illusion they have freedom, but actually they submit to an authority they're not aware of," Amanda told me. "They live to expectations anyways, instead of living for themselves."

I got the analogy. Whether she found physical freedom from her parents, system, or culture—America was certainly an ocean away from Shanghai—she was afraid she'd never find mental freedom.

"You will, Amanda," I told her. "You will find your own way. It takes time."

As the weeks counted down to her departure for college, one baby step toward liberty was prompted by her mother's own panic.

"They told me they regret what they did to me—raising me with so much pressure," Amanda told me, shaking her head. "She is a teacher, and my parents are more educated than most of the people in Shanghai. But they still did it to me anyway. Now? Now my mom is saying it's more important to be happy than successful."

To me, it was clear where blame should lie for each transgression in Amanda's young life: the insane pressure to perform on the Chinese system, the indestructible parent-child tether on Chinese culture.

But her mother's apology could only have been prompted by her illness.

* * *

FOR MY MASSEUSE FRIEND Lauren, both happiness and success seemed elusive.

Lauren's son, Little Jun Jun, would ultimately fail the High School Entrance Exam. I traveled out to Anhui province the weekend he sat for *zhongkao*, and as Lauren and I waited outside the gates, vultures were already circling: hired men passing out flyers to vocational schools. This was Plan B for students who failed to test into academic high schools on their merits. Lauren quietly took a flyer, folded it, and slotted it into her bag, while I pretended not to notice. Later that evening, back at her apartment, Lauren revealed yet another option. She'd been given the name and number of a high school principal the next county over. The man was known to take money for points, she muttered, eyes fixated on the pot of tofu she was tending for dinner.

A few weeks after I'd returned to Shanghai, I sent Lauren a message.

"How did he do?" I tapped.

My phone buzzed immediately with her reply:

> Jun Jun's candidate number is 139782900432. . . . His total score was 385.

Sitting in my living room in Shanghai, I gulped. "How much do you have to score to get into a normal high school?" I texted back.

"Five hundred," Lauren wrote. Jun Jun was so far beneath the cutoff line that a pay-for-points back door into a regular high school was likely out of reach.

Shortly thereafter, Lauren and her husband journeyed back to Shanghai, their business in the countryside finished. They considered bringing Jun Jun with them, but they couldn't bear to think of their eighteen-year-old son as a migrant laborer.

"But age eighteen is about when you went out to work for the first time," I reminded her, gently, when she came to see me the following fall. She wore a black top of see-through mesh and sequins, the flashy garb of a migrant in the big city, eager to show she had means.

"Yes, but this is my son," Lauren replied, glancing past my shoulder.

She'd enrolled Jun Jun at a vocational high school that took boarders back in Jingxian County. The school was a factory of sorts, of the type all too eager to take money from parents whose children had fallen off the academic track, in hopes they might still sit for the college entrance exam. The boy was now punching time against a three-year grind of six a.m. wake-up, twelve hours of daily study, and another two hours of review before lights-out—a *gaokao* exam-prep assembly line.

If his story could be told by statistics, he was unlikely to test into a top- or even second-tier college from this setup, but I also knew Lauren would be thrilled simply to delay the boy's entry into the migrant workforce for a few years. Tuition was 700 yuan a month.

"That's not bad!" I exclaimed.

"Yes," she said, "but there is a lot of competition for massages in Shanghai now. Many of my old clients have left the city."

"Patience," I told her, advocating the one quality no one in China seemed to appreciate. "Slowly, you'll make your way back."

Lauren glanced out my dining-room window. It was a good pollution day, and the sun's rays filtered into the room and glittered off the black sequins on her shirt.

"I have hope," she said.

* * *

MY TWO-YEAR-OLD SON, Landon, would be rejected at Soong Qing Ling.

"Blame me," I told Rob after it became clear I'd single-handedly gotten us into a deep pit of a mess, from which an admissions slip could never be extracted. I'd taken a stand on principle, and the school's authoritarian ways again opened up a hole underneath my feet.

The school offers a class for one- and two-year-olds called *qinziban*, or parent-child class. It's expensive—roughly $6,000 per year for a sound bite of a twice-weekly class—and it's widely viewed as a way for administrators to rake in extra cash. Many parents I knew paid the fees but either treated the class as social hour or attended only sporadically. There wasn't much of a direct benefit for the children.

In my mind, the setup was ethically questionable.

Unfortunately, Vice Principal Xi insisted I pay the parent-child fee to reserve a spot for Landon in the proper kindergarten, which begins taking kids the following year.

I rebelled. "I don't want to enroll him until he's three or four," I told Principal Xi. We'd found each other on Big Green at pickup time.

"You must pay for parent-child class to reserve a spot later," Xi repeated. At the time, Rob and I were dealing with Rainey's nap-time police threats, and I was in no frame of mind to commit my second, precious child to Soong Qing Ling.

Fast-forward a year or so. Rainey had adjusted, my heart and mind about the school coalesced, and we'd decided to pursue Chinese school early on in Landon's education, too. I approached Principal Xi to pick up the conversation.

Her eyes glinted. Uh-oh.

"You didn't pay the parent-child fee," she told me, parents swarming around her on Big Green. "I told you what would happen."

It would be an understatement to say that I begged and cajoled. I explained I didn't know her offer had an expiration date. I told Principal Xi that Landon had nowhere else to go. I proposed paying all back fees for the parent-child class. I enlisted a mutual friend to put in a good word. Still, the administration didn't relent, not even when I began bringing Landon with me to pick up his brother at the end of each school day. It was my feeble attempt at emotional manipulation, but I had little else to work with.

"Mommy, will this be my school, too?" Landon asked in his tiny voice, dirty blond hair ruffled and brown eyes wide, an exchange I conveyed to Principal Xi in an effort to appeal to her softer side.

Principal Xi would watch Landon toddle up to his older brother's classroom for pickup and totter back across Big Green, but her eyes seemed cold, and before long she began avoiding me. One week, the office admitted the son of American friends I'd counseled about the school. It felt like a slap in the face; our boys were the exact same age.

Landon would not attend Soong Qing Ling. Instead he'd begin at a bilingual international kindergarten, similar to the Victoria school we'd ultimately turned down for Rainey.

In the ensuing years, I'd wonder whether I should have capitulated early on to Principal Xi, especially as Landon struggled with the discipline and Chinese language learning for which Rainey's environment had laid smooth pavement. Landon's classroom—with one English-speaking and one Chinese teacher—had few of the academic habit-forming features and the high expectations of Soong Qing Ling.

On the other hand, Landon never asked permission to do anything.

⋆ ⋆ ⋆

ON THE LAST DAY of Middle Class grade, school let out a few hours early.

Rob and I chased after our son and a few hundred schoolmates as the children clambered down the stairs from the classrooms and hurtled onto Big Green. Bodies flooded the lawn, voices chirping at each other.

"*Fangxuele*—School's out!"

"*Dengwo*—Wait for me! Wait!"

"*Yiqi wanba!*—Let's play together over the summertime!"

Rainey linked hands with two classmates, Dongge and GuaZi, and the trio pulled each other toward the edge of Big Green and a row of bushes.

"Let's look for earthworms!" Rainey yelled, digging in a patch of chocolate-colored soil and pulling on the end of a wriggling pink mass of worm.

"Eggh," said Dongge, a pretty girl in an aquamarine dress, wrinkling her face. GuaZi also recoiled. Dongge's mother was a soft-spoken woman, a Buddhist, who never quite fit in with the regular crowd of fancy urbanites at Soong Qing Ling. Today she wanted to talk summer.

"Do you like to sail?" she asked, as our children frolicked near the bushes.

"We've never been sailing," I told her.

"Sailing is great," she said. "The children can learn balance and also be physically active at the same time." I'd not heard of sailing previously described in this manner.

A circus of excited children and parents raced toward the exit. The guards at the gate had changed out of their uniforms, and I detected some smiles among the phalanx of men that had once seemed so formidable. I glanced over at Rainey. Ever the patient leader, Rainey had managed to turn the girls' disgust into curiosity in under ten minutes flat, and Dongge was now holding worms up against the daylight, checking for transparency. A pile of wiggling masses colonized a small bed of leaves at her feet.

"Okay now, let's go ring the bell!" Rainey screamed out, changing course to clamber up the school's climbing trellis, twenty-foot-high yellow poles connected by horizontal ropes. A red metal bell hung at the top, and he grasped the rope and pulled with all his might, body leaning into thin air above our heads.

"*Ding ding ding! Dong dong dong!*" Rainey rang. "*Fangxuele*—School's out!" he screamed to his classmates, who continued to stream out the exit gates.

Dongge, her cotton dress billowing in the afternoon breeze, began scrambling up the rope, her face tilted upward. Soon, the two were ringing together.

"It's the last day of school!" Rainey shouted to the sky.

"It's the last day of school!" Dongge echoed, placing a hand just below Rainey's on the thick rope, their giggles mingling with the clanging chimes.

At ground level, Dongge's mother droned on about her morning routine: Mother and daughter listen daily to recordings of the writings of Confucius, Mencius, Laozi, and Zhuangzi.

"She's been doing this since she was three and a half," her mother said. "Research shows that the morning time after getting up is a golden time for the brain."

I thought about Rainey's morning routine: oatmeal and milk, followed by a half hour of Star Wars Lego play.

"Can Dongge *read* the classics also?" I asked, suspecting I already knew the answer.

"Yes, at night I'll sit down and read three passages each day. Then I'll point to each character as she recites from memory. She just picks it up. There are many beautiful sentences in those readings. Have you heard of the Book of Changes?"

"Yes, but I'm not familiar with the text," I replied.

I glanced up at Rainey and Dongge, one who enacted light-saber battles over breakfast, another who awoke each morning to Confucian chants. Tired of ringing together, they began taking turns yanking the rope. At this particular moment, it seemed nothing separated the two: not nationality, language, religion, or gender—not even proficiency with the recorder.

Dongge's mother wrapped up her pitch. "This week, we'll work on the *Classic of Filial Piety*," she said. "It doesn't matter how much you understand. She'll remember it after being exposed to it for a long time. You will love it—it's classically Chinese. You should try it with Rainey sometime," she said.

I glanced skyward. It was a good pollution day, and three bloated clouds swam lazily against long stretches of gray sky. The principals I'd tussled with for the last two years were nowhere in sight. The clanging from our happy children bounced off the skyscrapers towering over the school, echoing for a few moments, before dissipating into the ether.

"Maybe," I said. "Maybe I will."

ACKNOWLEDGMENTS

This is a work of nonfiction, resulting from seven years of living, reporting, and raising a family in Shanghai. I've changed names and identifying details where necessary to protect those who have divulged information or engaged in activities that would either compromise their privacy (or that of others) or draw scrutiny from the Chinese government. I've also changed the name of my son's school.

I'm grateful to the young Chinese who agreed to be part of this book. Darcy was a teenager when I first met him in Shanghai, and I watched him blossom into a sure-footed young adult who taught me how many Chinese reconcile what they're told with what is in their hearts. Amanda revealed to me the punishing mental sacrifices of a student in Shanghai; in her story I saw my younger self, and I find her resolve an inspiration. Lauren illuminated the brutal odds faced by rural families; may the winds bring her and her family everything they wish for. Thanks to all the others who consented to be interviewed for this book. This project would not have been possible without you. I also called on several part-time research assistants during the reporting and research phase, and I'm thankful for their dedication and trust. One chose to remain unnamed, and the others I'd like to mention are Michelle Hu, Shuang Wu, Yan Yuran, and Qingyi Zhu.

Thanks to Dorian Karchmar at William Morris Endeavor for her belief in me (and to Suzanne Gluck for helping me find her). I'm blessed to have not only a boxer for a literary agent but also a mentor in my corner. I'll always be grateful to Claire Wachtel for acquiring this project, and to Gail Winston at Harper for her masterly editing, and for shepherding the work onward with her sharp wisdom and steady hand. Thanks to Jonathan Burnham and Doug Jones for championing this book; to Sofia Groopman for her unflagging support; and to Heather Drucker, Nicole Dewey, Leah Wasielewski, Christine Choe, and the rest of the team for helping launch it into the world.

I benefitted from the advice, encouragement, and content expertise of colleagues and friends, including Alec Ash, James Areddy, Sebastien Carrier, Rebecca Catching, Margaret Conley, Stacy Cowley, Angela Doland, Clayton Dube, James and Deb Fallows, Russell Flannery, David Fleishman, Mei Fong, Jeremy Friedlein, Michelle Garnaut, Sig Gissler, Jeremy Goldkorn, York-Chi Harder and Stephen Harder, Peter Hessler, Corinne Hua, Vanessa Hua, Tina Kanagaratnam, Michael Kozuch, Kaiser Kuo, Margot Landman, Frank Langfitt, Jane Lanhee Lee, Ophelia Ma, Margo Melnicove, Adam Minter, Amanda Joan Mitchell, Michael Meyer, Crystyl Mo, Dipika Mukherjee, Phyllis Neufeld, Ching-Ching Ni, Evan Osnos, Sandy Padwe, Will Plummer, Qiu Xiaolong, Scott Rozelle, James Stigler, Peter Sweeney, Vijay Vatiheeswaran, Bruno van der Burg, Jeffrey Wasserstrom, Xinran, Jie Zhang, Sheen Zhang, and Michael Zilles. Steve Kettmann and Sarah Ringler hosted me at the Wellstone Center in the Redwoods, where I wrote three chapters beneath the Santa Cruz sunshine.

I'm indebted to friends who were willing to read all or parts of a draft and offer feedback, including Justin Bergman, Steven Bourne, Rachel Ee-Heilemann, Juthymas Harntha, Jessamine Koenig, Melissa Lam, Jen Lin Liu, Amy Poftak, Sarah Schafer, and Christine Tan. It's not easy getting that email with a 400-page document attached, and I'll always be thankful for your time. Thanks also to Jiang Xueqin for asking to see the manuscript, and special thanks to Victor Chiu for his meticulous eye for detail. I'm also obliged to Thomas Arnold, William Hogan, and Lucia Pierce for offering thoughts on the prologue.

A first book is an undertaking that requires audacity, mental resolve, and a physical stamina that calls for retreating into windowless rooms for months at a time. I rarely had these qualities in abundance, and never all at the same time, and when I wavered, my friends and loved ones cheered me on.

On that note, I'm thankful for the friendship and support, sometimes in the form of food, shelter, or babysitting, from the following: Alyshea Austern, Brantley Turner-Bradley and Doug Bradley, Alex Chen, Grace Chiu, Beth Colgan, Grace Lee and Dan Connelly, Fitz de Smet, Arsheya Devitre, Pete Dinh, Brenda Erickson, Eric Ericson, Christine Grand, Alexandra and Christian Hansmeyer, Steven Harris, Erich Heilemann, Ellen Himelfarb, Denise Huang, Heather and George Kaye, Lindsay Klump, Denise Landeros-Schmitz, Julie Langfitt, Rich Langone, Daniel Levine, Ann Meier, Maggie Moon, Kari Olson, Sonja Ortega, Caroline Pan, Lei Lei Peng, Jennifer Pitman, Michelle Rothoff, Ashley Schmitz, Dan Schmitz, Ryan Schmitz, Sophia Skoda, Alok Somani, Chris and Trask Stalnaker, Jason Stanard, Leslie Stephenshaw, Charles Tien, Liang Tien, Ming Tien, Mercedes Valle-Harris, Rishi Varma, Seth Werner, Stacey Wilson, Evonne Yang, and Rae Yang.

I couldn't have completed this project without the support of my family. I didn't have nearly enough time with my father-in-law, James Schmitz, but I'm grateful for the legacy he's left us and the boys. JoAnn Schmitz's calming influence and keen advice on everything from parenting to educating an elementary school child have been indispensable, and I loved talking about education with her closest teacher friends over red wine.

I'll always be grateful to my maternal grandparents, Tseng-Ying and Chin-Kai Tien; my fondest childhood memories come from visiting Grandpa's lab at the University of Michigan, and frolicking summers in that expansive backyard. To my paternal grandparents, Yung Chien

Chu and Wen-Yu Sun Chu, as much as I might complain about marathon cooking sessions for Chinese holidays, those are also some of my most cherished memories. Thank you all for helping me make my way.

My sister, Joyce Chu Moore, continues to amaze me every day, especially as she manages two toddlers in a brutal San Francisco Bay Area parenting scramble while also chairing a psychology department and running a training program to treat underserved minorities. She's a loyal friend and my most patient counselor, and I hope we'll someday live closer than an eleven-hour plane ride. I'm thrilled Kenney Moore is part of the family, and our doors are always open to him and to Greyson and Coralai.

While I may have protested their methods, I never argued with their motivations. My parents, Humbert and Judy Chu, taught me the value of hard work and schooled me with the attitude that anything is possible. Always outsiders in some sense, they were immigrants to America, and later we were one of the few Chinese American families in our neighborhood. I'll cherish every day of our journey, trying to figure it out together. Full disclosure: I disagree with my parents on the details of one anecdote. As I remember it, my mother and father refused to buy me a gerbil after I lost the piano competition described in these pages. Mom insists they carted their grumpy teenager off to the pet store anyway, where I stuck my finger into the gerbil cage, promptly got bitten, and forever swore off pet rodents. Whatever the truth, I always felt I was loved beyond the extent to which their arms could reach, and further than the depths of the heart. They poured their time and energy into ensuring my and my sister's future, many times at great sacrifice, and my own parenting efforts will always fall short.

In many ways, this book belongs to Rob Schmitz. Our life together has turned out to be an adventure I never could have imagined when we first struck up a conversation on the 1/9 train in New York. He has showed me the beauty in staring up at the stars and chasing a frog through the bushes, and taught me to seek a work life that's enjoyed rather than simply tolerated. I'm astounded by his talent for languages and for bringing stories to life, and I also aspire to his gift for seeing the light in any situation. I'm thrilled he's my life partner, and thankful that home feels like . . . home. My sons, Rainey and Landon, bring me more joy than I can describe, but I'll try. Rainey, you're an inspiration: I didn't have your resilience or your confidence until I was practically an adult. Thanks for dashing my fears about this project and diving in with enthusiasm: "I don't care, Mom! It's okay if people know things about me!" Landon, you bring light into my life and I'm sorry the office stole so much of my time these last couple of years. May the next stretch bring our family lots of world travel, Star Wars Lego sets, impromptu hunts for 黄鼠狼, and homemade chocolate chip cookies.

SELECTED BIBLIOGRAPHY

For my research, I conducted interviews with academics, Chinese parents, teachers, principals, and education policy makers, and experts. I consulted academic reports from China, the United States, Europe, and Australia; mainstream Chinese news media and online forums; and on occasion, Western news outlets. For data, I relied on Chinese government sources, including the Ministry of Education and the National Bureau of Statistics; Chinese state media; Chinese, American, and European researchers; and the work of not-for-profits and think tanks. What follows is a select bibliography of sources. I have translated the titles of Chinese sources into English. Generally, if the sources were referenced in more than one chapter, I placed them under General Sources. (For more detailed notes about facts, figures, and quotes, please visit my website www.lenorachu.com.)

PROLOGUE

Friedman, Thomas. "The Shanghai Secret," *New York Times*, October 23, 2013.

Levin, Richard. "The Rise of Asia's Universities." Speech delivered at the Royal Society, London, England, January 31, 2010, http://president.yale.edu/about/past-presidents/levin-speeches-archive.

Office of the Press Secretary, the White House. "Remarks by the President on the Economy in Winston-Salem, North Carolina." Posted on ObamaWhiteHouse .gov, December 6, 2010.

CHAPTER 1: FORCE-FED EGGS

Heng, Wang. "Cultural Interpretations of Socratic and Confucian Education Philosophy." PhD diss., University of Louisville, 2013
Shanghai Municipal Education Commission. "Notice on Further Strengthening the Work of Management and Selection of Municipal Exemplary Preschool." Office of Shanghai Municipal Education Commission, July 9, 2009.
Starr, Don. "China and the Confucian Education Model." Position paper commissioned by *Universitas* 21 (May 2012).

CHAPTER 2: A FAMILY AFFAIR

Interviews include Bo Weifeng, Ge Fang Ping, Gregory Yao, Ye Xue Jie.
Feng, Hairong, et al. "Examining Chinese Gift-giving Behavior from the Politeness Theory Perspective." *Asian Journal of Communication*, June 24, 2011.
Larsen, Janet. "Meat Consumption in China Now Double That in the United States." *Earth Policy Institute*, April 24, 2012.
National Bureau of Statistics. "2016 Migrant Work Monitoring and Research Report." National Bureau of Statistics website, April 28, 2017.
Zhu, Dongli. "The Structure and Social Function Analysis of Chinese Folk Present-Giving Custom." Northwest A & F University, 2010, 29–32.

CHAPTER 3: OBEY THE TEACHER

Interviews include Xu Song Gen, Huang Man Jie, Guo Li Ping.
National People's Congress. "Law of Compulsory Education of the People's Republic of China." April 12, 1986.
Soong Qing Ling School. *Child Development Book, 2013–2014*.

CHAPTER 4: NO EXCEPTIONS TO THE RULE

Interviews include Yang Qiao Yun, Zhou Xiao Lan.
Boroditsky, Lera. "Does Language Shape Thought?: Mandarin and English Speakers' Conceptions of Time." *Cognitive Psychology* 43 (2001), 1–22.
DeFrancis, John. *The Chinese Language: Fact and Fantasy*. University of Hawaii Press, 1984.

Environmental Protection Agency. *Air Quality Index: A Guide to Air Quality and Your Health* (February 2014).

Erbaugh, Mary S. "The Acquisition of Mandarin." *Crosslinguistic Study of Language Acquisition* 3 (1992), 373–455.

Juan, Yang. "Problems of Chinese Evaluation of Children with Special Needs." *Journal of Sichuan College of Education* 10 (2011).

Ministry of Education of the PRC. *Chinese Curriculum Standards of Compulsory Education (2011)*. Beijing Normal University Publishing Group, 2011.

———. "Twelve Departments' Advice on the Implementation of Standard Character List for Common Usage." Posted on the Ministry of Education website, October 15, 2013.

Ministry of Environmental Protection of the PRC. "List of Enterprises under Key Supervision of the Nation in 2016." February 4, 2016.

Moser, David. "Why Chinese Is So Damn Hard." *Schrifestschrift: Essays on Writing and Language in Honor of John DeFrancis on His Eightieth Birthday*. Sino-Platonic Papers, 1991.

Shanghai Board of Education. "Heavy Air Pollution in Shanghai, Education Commission Requires Suspension of All Outdoor Activities for Students." *People's Daily* website. December 2, 2013.

Twenty-first Century Institute of Education. *2014 Blue Book of Education*. Social Sciences Academic Press, 2014.

Wong, Edward. "'Airpocalypse' Smog Hits Beijing at Dangerous Levels." *Sinosphere* (blog), *New York Times*, January 17, 2014.

CHAPTER 5: NO REWARDS FOR SECOND PLACE

China Education Online. "The History of the Gaokao." Posted on ifeng.com, April 21, 2010.

China News Network. "In 2014, 7.97 Million Students Were Admitted into Academic High Schools, 260 000 Less than Last Year." July 30, 2015.

Civil Service Examinations, Berkshire Encyclopedia of China. Berkshire Publishing Group 2009.

Crozier, Justin. "Chinese Imperial Examination System." *China in Focus* (2002).

Decisions of the National Language Committee, the National Education Committee and Ministry of Radio, Film and Television on the Test of Mandarin Level. October 30, 1994.

Elman, Benjamin A. "Political, Social, and Cultural Reproduction via Civil Service Examinations in Late Imperial China." *Journal of Asian Studies* 50:1 (1991), 7–28.

Fan Guangji. "The Historical Evolution of China's College Entrance Examination

System and Its Revelation to the Reform of Modern College Entrance Examinations." Fujian Education College Newspaper (June 2011).

Liu Fang. "Students Admitted to Vocational Schools Occupied 44.12% of the Total Number in the High School Recruitment Period in 2014." September 16, 2015.

Min Huang. "Xiaogan High School Put 12th Grade Students on a Drip to Supplement Energy." *Changjiang Times*, June 6, 2012.

Ministry of Education of the PRC. "Statistics on the Number of Gaokao Participants and Those Admitted over the Years." Posted on *People's Daily* online, May 3, 2013.

———. "Overview of Chinese Education." November 25, 2015.

Miyazaki, Ichisada. *China's Examination Hell: The Civil Service Examinations of Imperial China*. Yale University Press, 1981.

Sina Education. "Ranking of *Yiben* Admission Rate of All Provinces, Beijing, Tianjin and Shanghai Top the Chart." Posted on Sina.com, December 25, 2015.

Song Chunpeng. "The Double-Edged Sword of Traditional Culture and Scientific Innovations—Some Thoughts on the Talent Development of Universities." Posted on the Ministry of Education of the PRC website, April 24, 2008.

Suen, H. K. "The Hidden Cost of Education Fever: Consequences of the Keju-Driven Education Fever in Ancient China." Translated by Ki-soo Kim. Hawoo Publishing Co., 2005.

ww123 Forum. "Parent in Shanghai." April 23, 2014.

Xinhua News Agency. "In 2014 the Country 9.39 Million Candidates to Participate in College Entrance Examination." Posted on moe.gov.cn, June 4, 2014.

CHAPTER 6: THE HIGH PRICE OF TESTS

Interviews include Cong Qianzhen, Kang Jian, Li Jiacheng, Liu Jian, Li Qiong, Tiehui Weng, Wang Feng, Xu Jiandong, Yang Dongping, Yang Linqiong.

Chai Wei and Xiang Qianyun, "China's 'PISA'—Unveiling China's Educational Assessment System." Posted on *China Education Daily*, April 16, 2015.

Chongqing Morning News. "Chongmin Luo Delivered Speech in Southwest University: Last Year in Middle and High School Waste a Lot of Time." Posted on Xinhuanet.com, May 5, 2012.

Du, Feijin et al. "Heavy Burden on Students Has Become the Pain of the Nation." *People's Daily*, August 2, 2013.

Dukes, Richard, and Heather Albanesi. "Seeing Red: Quality of an Essay, Color of the Grading Pen, and Student Reactions to the Grading Process." *Social Science Journal* 50 (2013), 96–100.

Feng Jianjun. "Examination-Oriented Education Wears Down Student's Personality." Posted on China Education News Network website, April 21, 2015;

Guangming Daily. "Primary and Middle School 'Jianfu' Report Published." Posted on *Xinhua News* online, May 31, 2013.

Gu Mingyuan. "Interest Is the Biggest Drive of Study." Posted on qq.com, August 15, 2014, http://edu.qq.com/a/20140815/015290.htm.

He Yingchun. "Reflection on 'Fudan Poisoning Case': We Need to Learn to Forgive." Posted on people.com, February 18, 2014.

Ji Lin. "Wei Yu: Socio-Emotional Competency Will Affect Children for the Lifetime," *Shanghai Education News*, April 28, 2013.

Liang Jianmin, et al. "Zheng Yefu: Primary and Middle School Education Is Killing Creativity." *Information Times*, October 9, 2013.

Li Jimei and Feng Xiaoxia. *Interpretation of Guideline of Children's Learning and Development for 3- to 6-year-old Children*, People's Education Press, 2013.

Liu Shiyu. "Ni Minjing: The Meaning of PISA Test Is to Enlighten the Evaluation Reform." *Shanghai Education News*, December 19, 2013.

Luo Yangjia, Shanghai Education. "Shanghai Student's Top Performance in PISA Evokes Thinking." Posted on Xinhuanet.com, March 17, 2011.

Ma Bei Bei. "National Physique Evaluation: Chinese Youth's Physique Has Decreased for 10 Years." *China Youth Daily*, March 30, 2010.

Ministry of Education of the PRC, Department of Basic Education. *Interpretation of Kindergarten Education Guidelines*. Phoenix Education Publishing, 2002, repr. 2012.

———. "Guidance on Current Efforts to Strengthen the Management of Primary and Secondary Schools." April 22, 2009.

———. "National Outline for Medium and Long-term Education Reform and Development (2010-2020)." July 2010.

———. "On the Prevention and Correction of Preschool 'Elementary-Schoolization.'" December 28, 2011.

———. *Guideline of Children's Learning and Development for 3- to 6-year-old Children*. Capital Normal University Press, 2012.

———. "To Solicit Public Opinion on 'Ten Rules on Burden Relief of Elementary School Students.'" August 22, 2013.

———. "Notice of the Issue of Guide of the Establishment of Psychological Room in Middle and High Schools." July 31, 2015.

National People's Congress (NPC). "Education Law of the People's Republic of China." Posted on Chinese government portal gov.cn, May 25, 2005.

Netease Education Channel. "Research of Study Pressure of Chinese Primary and Middle School Students." November 25, 2015.

Play Mountain Place website. www.playmountain.org/.

Pleasance, Chris. "NY Teacher Brands Common Core Reform Child Abuse, Gets Rapturous Applause." *Daily Mail*, November 15, 2013.

Preschool to Primary School Web. "Beijing 2014 Primary School Entrance Exam Collection: Common Sense." Posted on ysxiao.cn, October 28, 2014.

Quality Education Research Group. "Common Concern: Quality Education System Research." Educational Science Publishing House, 2006.

Ren Jinwen. "Hainan Gaokao Reform Breaks the System of Marks, Gives the Universities Autonomy for Admission." *Haikou Evening News*. Posted on Xinhuanet.com, August 5, 2014.

Ryan, Janette. *Education Reform in China: Changing Concepts, Contexts, and Practices*. Routledge, 2011.

Salinas, D. "Does Homework Perpetuate Inequities in Education." *PISA in Focus* 12 (2014), 1–4.

Shanghai Municipal Education Commission. "Several Opinions about Effectively Alleviating the Academic Burden on Schoolchildren." December 5, 2004.

———. "The Notice of the Issue of the Pilot Plan of Shanghai Spring Gaokao in 2017." November 8, 2016.

Shanghai Star. "Five Bears Suffer Sulphuric Acid Attack." Posted on *Shanghai Star* website, February 28, 2002.

Sohu.com. "Shanghai Universities Will Consider Normal High School Test, Students Will Select Three Subjects Themselves." *Sohu*, January 7, 2016.

Soong Qing Ling Primary School Readiness Meeting Notes, April 22, 2014.

Southern Weekly. "Top Controversy: Should the Act of Good Samaritan Justify the Recommendation Admission to the University?" Posted on ifeng.com, June 13, 2014.

State Council of the CP CCC. "China Education Reform and Development Guidelines." Posted on Ministry of Education of the PRC website, February 13, 1993.

———. "Decision on the Deepening of Education Reform and the Enhancement of Suzhi Education." General Office of the CPC Central Committee, June 13, 1999.

———. "The Implementation of Performance Pay Compulsory School Guidance." Legislative Affairs Office of the State Council, December 23, 2008.

———. "Advice on Deepening the Examination and Admission System Reform." September 4, 2014.

———. "Full List of Practice Certifications for Different Jobs." Posted on State Council website, December 19, 2016.

State Education Committee of PRC. "Some Notes about Actively Promoting Quality-Oriented Education." 1997.

Tencent Education. "21st Century Education Research Institute Compulsory Education Stage of Primary and Secondary School Students 'Lessen the Burden' Situation Survey." 21st Century Education Research Institute, May 31, 2013.

Wang Wei. "PISA Only Tests One Aspect of the Education." *Shanghai Xinmin Evening News*, December 5, 2013.

———. "Shanghai May Quit the Next PISA Test." *Shanghai Xinmin Evening News*, March 7, 2014.

Wang Xiao Li. "Thoughts on the Change of Burden Relief Policies on Basic Education since the Foundation of the PRC." *Education and Examinations* 5 (2009).

Wang Ying. "9.39 Million Students Took Gaokao in 2014 Nationwide." Posted on Xinhuanet.com, June 4, 2014.

Xie Yu et al. "The Report on Civic Development in China 2013." Peking University Press, September 1, 2013.

Zhai Zihe. "The Fudan Poisoning Case Sensationalized the Society: Roommates Farewell the Dead in Tears." *Xinhua News*, April 17, 2013.

Zhang Minxuan. "What Caused the Misinterpretation of the PISA Result?" *Wenhui Daily*. December 16, 2013.

Zhi Zhi and Huang Yiming. "Student Killer an Introvert Who Finally Cracks." *China Daily*, March 17, 2004.

CHAPTER 7: LITTLE SOLDIER

Interviews include Leland Cogan, Tan Chuanbo, Wang Zheng, William Schmidt, Xie Xiaoqing, Yong Zhao, Zhao Jue.

Bakken, Børge. *The Exemplary Society: Human Improvement, Social Control, and the Dangers of Modernity in China*. Oxford University Press, 2000.

Barboza, David. "Billions in Hidden Riches for Family of Chinese Leader." *New York Times*, October 25, 2012.

Cheung Kwah, Pan S. "Transition of Moral Education in China: Towards Regulated Individualism." *Citizenship Teaching and Learning* 2:2 (2006), 37–50.

China Network. "Evolution of Primary School Textbook over the Past 60 Years." Posted on China.com.cn, July 11, 2009.

Communist Party of China. "Rules of the Communist Party of China on the Development of Communist Party Members." Posted on cpcnews.cn, undated.

———. "Relationship between the Communist Party and the Communist Youth League." Posted on gqt.org.cn, December 29, 2012.

———. "Seminar on the Implementation of the Document No. 9 Issued by the General Office of CPC Central Committee," Posted on the Shanghai University of Political Science and Law website, January 23, 2014.

Dai Weisen. "Hong Kong Education System Should Be More Patriotic: Chinese Officials." Posted on Radio Free Asia website, January 8, 2015.

Darcy. "Walking on the Red Road" letter. Study process at Party school, accessed April 12, 2014.

Fifth National People's Congress. "The Regulation of the Young Pioneers of China." June 3, 2005.

Forsythe, Michael, and Shai Oster. "Xi Jinping Millionaire Relations Reveal Elite Chinese Fortunes." Bloomberg News, June 29, 2012.

Grammes, Tilman. "Nationalism, Patriotism, Citizenship and Beyond: Questioning the Citizenship Industry." *Journal of Social Science Education* 10:1 (2011), 2–11.

Implementation Outline of Patriotism Education. People's Education Press, 1994.

Jacobs, Andrew. "Chinese Professor Who Advocated Free Speech Is Fired." *New York Times*, December 10, 2013.

Johnson, Ian. "Learning How to Argue: An Interview with Ran Yunfei." *New York Review of Books*, March 2, 2012.

Jin Hao. "Up to the End of 2015, There Are 88 Million of League Members." *Guangming Daily*, May 4, 2016.

King, Gary, et al. "Reverse-Engineering Censorship in China: Randomized Experimentation and Participant Observation." *Science* 6199 (2014), 345: 1–10.

Kipnis, Andrew B. *Governing Educational Desire: Culture, Politics, and Schooling in China*. University of Chicago Press, 2011.

Kondo, Takahiro, and Xiaoyan Wu. "Patriotism as a Goal of School Education in China and Japan." *Journal of Social Science Education* 10 (Spring 2011), 1.

Lee, Wing On, and Chi Hang Ho. "Ideopolitical Shifts and Changes in Moral Education Policy in China." *Journal of Moral Education* 34:4 (2005), 413–31.

Li Lihong. "Promote 'Big Youth League Committee,' Welcome the 18th National Congress of the CPC." *China Youth News*, October 10, 2010.

Maosen, Li. "Moral Education in the People's Republic of China." *Journal of Moral Education* 19 (1990), 3.

———. "Changing Ideological-Political Orientations in Chinese Moral Education: Some Personal and Professional Reflections." *Journal of Moral Education* 40:3 (2011), 387–95.

Mingjing News. "Gao Yu in Leak Case Sentenced to Seven Years." Posted on chinaaid.net, April 17, 2015.

Phillips, Tom. "'It's Getting Worse': China's Liberal Academics Fear Growing Censorship." *Guardian*. August 6, 2015.

Plan of Nine-Year Compulsory Full-Time Primary School Curriculum and Middle School Curriculum. People's Education Press, 1992.

"Practice and Reflection of Developing High School Student Party Members." *Ideological and Theoretical Education* 16 (2009), 1.

Propaganda Department of the Central Committee of the CPC. "Patriotism Education Implementation Summary." *People's Education* 10 (1994), 6–9.

Ran Zhang. "By the End of 2015, There Are 88.758 Million Party Members in China." *Jing Hua*, July 1, 2016. Reposted on People.com.

Seventh National People's Congress. "The Law of the People's Republic of China on the National Flag." October 1,1990.

Shepherd, Christian. "China Changes Start Date of War with Japan, Says Will Bolster Patriotic Education." Reuters, January 11, 2017.

Tsang, Mun. *Education and National Development in China Since 1949: Oscillating Policies and Enduring Dilemmas*. China Review, 2000.

Wang, Zheng. "National Humiliation, History Education, and the Politics of Historical Memory: Patriotic Education Campaign in China." *International Studies Quarterly* 52 (2008), 783–806.

Wansheng, Zhan, and Ning Wujie. "The Moral Education Curriculum for Junior High Schools in 21st Century China." *Journal of Moral Education* 33 (2004), 4.

Westheimer, Joel. "Should Social Studies Be Patriotic?" *Social Education* 73:7 (2009), 316–20.

Writing Group of High School Ideology and Politics. *Textbook of MOE, Ideology and Politics*. People's Education Press, 2013.

Xinhua News Agency. "Authorized Release: The CPC's Statistical Communique in 2015." Posted on Xinhuanet.com, June 30, 2016.

Zang Xiaowei. "Educational Credentials, Élite Dualism, and Élite Stratification in China." *Sociological Perspectives* 44:2 (2001) 189–205.

Zhao, Yong. *Who's Afraid of the Big Bad Dragon*. Jossey-Bass, 2014.

Zhao Wei. "The First Chinese Lesson." *Procuratorial Daily*, August 28, 2015.

CHAPTER 8: ONE HUNDRED DAYS 'TIL TEST TIME

Interviews include Guan Yi, Scott Rozelle, Zhang Futao, Zhou Nian Li.

Chan, Kam Wing. "The Household Registration System and Migrant Labor in China: Notes on a Debate." *Population and Development Review* 36:2 (2010), 357–64.

———. "China, Internal Migration." *Encyclopedia of Global Migration*. Blackwell Publishing, 2013.

"China's First Poverty-Alleviation Day. a List of 592 Key Counties of National Poverty Alleviation and Development." October 17, 2014.

Dahe.cn. "216 Students in Henan Province Have Been Enrolled by Peking University. about 70% of Gaokao Champions in 18 Cities Have Chosen Peking University. Posted on Dahe.cn, July 21,2016.

The Economist. "Ending Apartheid, Special Report." *Economist*. April 19, 2014.

———. "Kaifeng: Down and out in rural China." *Economist,* August 23, 2014.

Li Shi. "Speech about the Gini Coefficient at the China Economic Development Forum 2016." Posted on Sina.com, November 19, 2016.

Liu Jinsong. "Peking University Professor: Guangdong and Anhui Students' Rate of Admission in Peking University Is Only 1 Percent That of Beijing Students." *Economic Observer*, June 1, 2012.

Liu Yaxuan. "The Patriotism in the Textbook of History in Foreign Primary Schools." *Journal of Teaching and Management*, May 15, 2012.

Mao Zedong. *On Agriculture Cooperation.* From *Selected Works of Mao Zedong.* People's Publishing House, 1977: 168–91.

Ministry of Education of the PRC, Department of Development Planning. *China Education Yearbook 2010.* People's Education Press, 2010.

Pepper, Suzanne. *Radicalism and Education Reform in Twentieth-Century China.* Cambridge University Press, 1996.

Peterson, Glen. *The Power of Words: Literacy and Revolution in South China, 1949–1995.* University of British Columbia Press, 1997.

Roberts, Dexter. "Chinese Education: The Truth Behind the Boasts." Bloomberg News, April 5, 2013.

Rose. "Target Hackathon: Design an App for China Left-behind Children." Posted on Collective Responsibility, 2017, http://www.coresponsibility.com/target-hackathon-left-behind-children/.

Ross, Heidi. *China Country Study.* Paper commissioned for the EFA Global Monitoring Report, *2006, Literacy for Life*, 2005.

Rozelle, Scott. "Caixin Column 2: China's Inequality Starts during the First 1,000 Days." *Caixin* magazine series *Inequality 2030* (October 28 2013).

Seeborg, Michael C., et al. "The New Rural-Urban Labor Mobility in China: Causes and Implications." *Journal of Socio-Economics* 29 (2000), 39–56.

Shirk, Susan L. "Work Experience in Chinese Education." *Comparative Education* 14 (1978), 5–18.

"Tabulation on the 2010 Population Census of The People's Republic of China." China Statistics Press.

21 Century Education Research Institute. "Civil Gaokao Reform Scheme Version 3.0." Posted on the 21 Century Education Research Institute website, 2014.

Wu Ce, Pang Zheng. "National Poverty Relief Priority Counties List Released." Posted on China.com, June 8, 2016.

Wu Han Evening Newspaper. "The Proportion of Rural College Students Causes Wen Jiabao Concern." Posted on Sina.com, January 24, 2009.

Yuan Guiren. "Comprehensively Promote the Scientific Development of National Education Cause." Posted on MOE website, January 3, 2011.

Zhao, Yaohui. "Labor Migration and Earnings Differences: The Case of Rural China." *Economic Development and Cultural Change* 47:4 (1999), 767–82.

Zhen Yang. "9.02 Million! Latest National Rural Left-behind Children Statistics Published—Why Has the Number Dropped 50 Million?" Posted on Kanka news, November 9, 2012.

CHAPTER 9: SHORTCUTS AND FAVORS

Baidu Online Forum. "The Initial Gaokao Papers Were Leaked in Hunan 2015." Posted anonymously, June 24, 2015.

Bain & Company. "The Rise of the Borderless Consumer." *Luxury Goods Worldwide Market Study*, Fall–Winter 2014.

BBS Online Forum. Post by teacher in Ezhou City, April 17, 2012.

Central Commission for Discipline Inspection. "Ministry of Education: Paid Afterschool Classes Banned and Hotline for Complaints Published." July 4, 2015.

China News Network. "Henan Investigation of College Entrance Examination for the 127 People." June 18, 2014.

———. "Peking University and Tsinghua University Plan to Enroll 353 Students in Beijing 2016." Reposted on Sina.com, June 20, 2016.

Fang Zhou. "Can We Have No-Gift Teachers' Day, Please." ChinaDaily.com, September 10, 2014.

Fan Tianjiao. "Several Officials of Education Sector Dismissed for Bribery in Bengbu, Education Equipment Procurement Section Becomes Easy Target for Bribery." *Legal Daily*, October 12, 2016.

Gough, H. G. "A Creative Personality Scale for the Adjective Check List." *Journal of Personality and Social Psychology* 37:8 (1979), 1398.

Gu Yuansen. "Cai Rongsheng, Former Head of Admission and Employment at Renmin University, Was Accused." *Modern Express*, October 10, 2015.

"Heat News: Go Undercover in a Cross Province Group Providing Substitute with the Examinee in Gaokao." June 7, 2015. Reposted on Sina.com.

Hou, Ning, et al. "Survey and Regulation on Cheating Behavior of College Students." *Examination Journal* 73 (2011).

Jou, Eric. "Tool for Cheating in China Is Comparable to 007's." *Reference for the Youth*, June 25, 2014.

Liu, Jintie, et al. "Investigation on University Students' Cheating in Examinations and Corresponding Countermeasures." *Journal of Liaoning Technical University* (Social Science Edition) 14 (2012), 1.

Moore, Malcolm. "Riot after Chinese Teachers Try to Stop Pupils Cheating." *Telegraph*, June 20, 2013.

Ministry of Education of the PRC. "The Notice of the Issue of Regulations on Prohibiting Teachers Accepting Gifts or Payments from Students and Their Parents by Ministry of Education." July 8, 2014.

———. "Establishing a Sound System of Punishing and Preventing Corruption in Organs and Units Directly Supervised by the Ministry of Education." Ministry of Education of the PRC, October 14, 2014.

———. "Regulations on the Prohibition of School and Teachers Employed to Give After-Class Lessons." Ministry of Education of the PRC, June 29, 2015.

———. "Anti-Corruption in Education System Is Still Tough and Complex." Posted on Paper.cn, February 27, 2016.

———. "Special Campaign by Ministry of Education to Address the Problem of Paid Make-up Class and Teachers Accepting Gifts and Payments against Regulations." July 26, 2016.

Ministry of Finance of the PRC. "The Implementation Advice on Regulating the Collection of Fees for Education and Rectifying Unauthorized Collection of Fees for Education in 2015." June 3, 2015.

National People's Congress of the People's Republic of China. "The 163rd Clause of the Criminal Law of the People's Republic of China." 2015.

———. "Version 9. Amendment to Criminal Law of the People's Republic of China." August 29, 2015.

Nelson, Katie. "Photos of the Day: Hubei School Holds Midterms in Forest to Prevent Cheating." *Shanghaiist*, May 3, 2014.

Ni Zhang. "How Strict Is Gao Kao Exam Discipline This Year?" China News Network, May 25, 2016.

Shanghai Sheng Xue. "Shanghai High School Ranking Based on Enrolled Number of Students into Local Yiben Universities." Posted on edu.online.sh.cn, July 8, 2016.

Strait News. "One Student in Fujian Paid for a Surrogate Exam-Taker for Gaokao; Both Were Jailed." July 28, 2016.

Westwood, R., and D. R. Low. "The Multicultural Muse: Culture, Creativity and Innovation." *International Journal of Cross-Cultural Management* 3:2 (2003), 235–59.

Wu Nan. "Kindergarten Drug Scandal Leads to Call for Overhaul of Regulations." *South China Morning Post*, March 30, 2014.

Xinhua News Agency. "Amendments to the Criminal Law of the People's Republic of China." August 30, 2015.

Yang Meiping. "Teachers Seduced by Lucrative Private Work." *Shanghai Daily*, February 22, 2016.

Zahedi, F. "The Challenge of Truth Telling across Cultures: a Case Study." *Journal of Medical Ethics and History of Medicine* 4 (2011), 11.

Zhu Dongli. "The Structure and Social Function Analysis of Chinese Folk Present-Giving Custom." Northwest A & F University, 2010, 29–32.

CHAPTER 10: BEATING THE SYSTEM VERSUS OPTING OUT

Interview subjects include Carol Dweck, Chai Bensheng, Charles Leadbetter, Haiyan Hua, Marc Tucker, Xinran.

CCTV-2. "Dialogue: Ballpoint Pen Challenges High End Manufacture." November 22, 2015.

Didau, David. "PISA 2015: Some Tentative Thoughts about Successful Teaching." *The Learning Spy* blog, December 6, 2016.

Dong Hongliang. "Yuan Guiren on Education Dream: Individualized Education for All." *People's Daily*, March 8, 2013.

Fu, Genyue, et al. "Chinese Children's Moral Evaluation of Lies and Truths: Roles of Context and Parental Individualism-Collectivism Tendencies." *Infant and Child Development* 19:5 (2010), 498–515.

Guilford, J. P. "Creativity." *American Psychologist* 5:9 (September 1950), 444–54.

Han Han. "Shower with the Quilted Jacket On." Posted on Sina.com, September 29, 2009.

Kim, Kyung Hee. "Learning from Each Other: Creativity in East Asian and American Education." *Creativity Research Journal* 17 (2005) 4: 337–47.

Liu Chang. "Chinese Company Successfully Manufacturers Ballpoint." Posted on cctv.com, January 10, 2017.

Li Xiying, et al. "The Relationship of Youth's Creative Personality and Creativity: A Comparative Study between China and U.S." *Psychological Exploration* 34 (2014) 2: 186–92.

Postiglione, Gerard. "Improving Transitions: From School to University to Workplace." Asian Development Bank, 2012.

Robinson, Ken. *Creative Schools: The Grassroots Revolution That's Transforming Education*. Penguin Publishing Group, 2016.

SASAC. "China Aerospace Science and Industry: Fully Tap into the Role Creativity Plays in Corporate's Development." Posted on 163.com, December 7, 2012.

Schmitz, Rob. "Why Can't China Make a Good Ballpoint Pen?" *APM's Marketplace*, December 14, 2015.

Schwinger, Malte, et al. "Academic Self-Handicapping and Achievement: A Meta-Analysis." *Journal of Educational Psychology* 106:3 (2014), 744–61.

Shen Jiliang. "The Cultivation of Creative Talents in China Through Intercultural Comparison." *Chinese Talents* 11 (2014).

Staats, Lorin K. "The Cultivation of Creativity in the Chinese Culture—Past, Present, and Future." *Journal of Strategic Leadership* 3:1 (2011), 45–53.

State Council of the CPCCC. "Advice on Policies That Promote Popular Entrepreneurship and Innovation." July 16, 2015.

———. "The 13th Five Year Plan for National Education Development." January 19, 2017.

Wadhwa, Vivek, et al. "America's New Immigrant Entrepreneurs, Parts I–VII." Duke University, Master of Engineering Management Program; University of California–Berkeley, School of Information, January 4, 2007.

———. "Globalization of Innovation." Kauffman Foundation of Entrepreneurship, June 2008.

——. "Skilled Immigration and Economic Growth." *Applied Research in Economic Development* 5 (2008): 1.

Xia Guming. "Research on Students' Critical Thinking," Posted on People Education Press, February 7, 2012;

Xiao Gangling. "Why Not Educate Students to 'Challenge Authority' during Conventional Teaching?" China Education News Network, May 14, 2015.

Yangguang News. "Li Keqiang Gave an Important Delivery on the Symposium on Reform and Innovation of Higher Education." April 18, 2016.

Yun Ma. "If I Graduated from Tsinghua or Peking University . . ." Zhejiang Chamber of Commerce in Shanghai, ninth member meeting. Posted on Sina.com, December 8, 2014.

Zheng, Yefu. *The Pathology of Chinese Education*. CITIC Press, 2013.

CHAPTER 11: LET'S DO MATH!

Interviews include Jim Cocoros, David Didau, Andreas Schleicher, Brian Steuter, James Stigler, Jie Zhang, Yang Xiaowei, Zheng Zhou.

Benchmarking for Success. *Report by the National Governors Association*. Council of Chief State School Officers and Achieve, Inc.

Center on International Education Benchmarking. "9 Building Blocks for a World-Class Education System." ncee.org, December 2016.

China Youth Daily. "British Visit Shanghai to Learn How to Teach Mathematics." Posted on edu.com, February 28, 2014.

Crawford, Claire, and Jonathan Cribb. "Reading and Maths Skills at Age 10 and Earnings in Later Life: A Brief Analysis Using the British Cohort Study." *Institute for Fiscal Studies,* March 8, 2013.

Duncan, Greg. "School Readiness and Later Achievement." *Developmental Psychology* 43:6 (2007), 1428–46.

Education Online. "The Number of People Studying Abroad from 2006 to 2012." Report of overseas study. Posted on Education Online, 2013.

Fuller-Rowell, T., and S. Doan. "The Social Costs of Academic Success across Ethnic Groups." *Child Development* 81:6 (2010), 1696–1713.

Grant, Leslie, et al. *West Meets East: Best Practices from Expert Teachers in the U.S. and China*. ASCD, 2014.

Huntsinger C. S, et al. "Cultural Differences in Early Mathematics Learning: A Comparison of Euro-American, Chinese-American, and Taiwan-Chinese Families." *International Journal of Behavioral Development* 21:2 (1997), 371–88.

Jerrim, John. "The Link between East Asian 'Mastery' Teaching Methods and English Children's Mathematics Skills." *Economics of Education Review* 50 (2016), 29.

Kirschner, Paul A. "Why Minimal Guidance during Instruction Does Not Work: An Analysis of the Failure of Constructivist, Discovery, Problem-Based, Experiential, and Inquiry-Based Teaching." *Educational Psychologist* 41:2, 75–86.

Klahr, D., and M. Nigam. "The Equivalence of Learning Paths in Early Science Instruction: Effects of Direct Instruction and Discovery Learning." *Psychological Science* 15:10 (2004), 661–67.

Ma, Liping. *Knowing and Teaching Elementary Mathematics: Teachers' Understanding of Fundamental Mathematics in the United States and China.* Routledge, 2010.

Ministry of Science and Technology. "Outline of the National Plan for the Development of Science and Technology for 1978–1985 (Draft)." Posted on the Ministry of Science and Technology website. December 1977.

Morgan, Paul, et al. "Which Instructional Practices Most Help First-Grade Students with and without Mathematics Difficulties?" *Educational Evaluation and Policy Analysis* (2014).

Niu, W., and Z. Zhou. "Teaching Mathematics Creativity in Chinese Classrooms." In *Nurturing Creativity in the Classroom*. Cambridge University Press, 2010.

Peterson, Paul E., et al. "Globally Challenged: Are U.S. Students Ready to Compete?" Prepared under the auspices of Harvard's Program on Education Policy and Governance, and Education Next, August 2011.

Ryan, Julia. "American Schools vs. the World: Expensive, Unequal, Bad at Math." *Atlantic*, December 10, 2013.

Shellenbarger, Sue. "The Best Language for Math: Confusing English Number Words Are Linked to Weaker Skills." *Wall Street Journal*, September 15, 2014.

Siegler, R. S., and Y. Mu. "Chinese Children Excel on Novel Mathematics Problems Even before Elementary School." *Psychological Science* 19:8 (2008), 759–63.

Stevenson, Harold, James W. Stigler. *Learning Gap: Why Our Schools Are Failing and What We Can Learn from Japanese and Chinese Education.* Summit Books, 1992.

Sztein, Ester, and National Research Council. *The Teacher Development Continuum in the United States and China: Summary of a Workshop*. National Academies Press, 2010.

Vigdor, Jacob. "Solving America's Math Problem." *EducationNext* 13 (2013), 1.

Winifred, Mark, and Ann Dowker. "Linguistic Influence on Mathematical Development Is Specific Rather than Pervasive: Revisiting the Chinese Number Advantage in Chinese and English Children." *Frontiers in Psychology* 6 (2015), 203.

Zhao, D., and M. Singh. "Why Do Chinese-Australian Students Outperform Their Australian Peers in Mathematics: a Comparative Case Study." *International Journal of Science and Math Education* 9 (2011), 69.

Zhou, Zheng, et al. "Understanding Early Mathematical Competencies in American and Chinese Children." *School Psychology International* 42:3 (2005), 259–72.

Zhou, Zheng, and Stephen Peverly. "Teaching Addition and Subtraction to First Graders: A Chinese Perspective." *Psychology in the Schools* 42 (2005), 3.

CHAPTER 12: GENIUS MEANS STRUGGLE

Interviews include Zhang Zhizhong, Sheen Zhang, Teacher Mao, Teacher Song, Xiaodong Lin, Zhang Huiping.

Alexander, Scott. "The Parable of the Talents." *Slate Star Codex* blog, January 31, 2015.

Bjork, Robert A. "Memory and Metamemory Considerations in the Training of Human Beings." *Metacognition: Knowing About Knowing.* MIT Press, 1994.

Borden, Jonathan. *Confucius Meets Piaget: An Educational Perspective on Ethnic Korean Children and Their Parents.* 2nd edition. Seoul Foreign School, 2003.

Dolton, Peter, and Oscar Marcenaro-Gutierrez. *Global Teacher Status Index.* Varkey GEMS Foundation, 2013.

Dweck, Carol S. "Caution—Praise Can Be Dangerous." *American Educator* (Spring 1999).

Gong Shanshan, "The Positive and Negative Effects of Rewards and Punishment in Education and Proper Ways of Using Them." *Studies in Logic* 26 (2006), 190–93.

Heffernan, Virginia. "Drill, Baby, Drill." *New York Times Magazine*, September 16, 2010.

Hirsch, Ed. "You Can Always Look It Up, or Can You?" *American Educator* (Spring 2000).

Hong, H.-Y., and X. Lin-Siegler. "How Learning about Scientists' Struggles Influences Students' Interest and Learning in Physics." *Journal of Educational Psychology*. First published online November 11, 2011.

Hsin, Amy, and Yu Xie. "Explaining Asian Americans' Academic Advantage over Whites." *Proceedings of the National Academy of Sciences* 111 (2014), 23.

Maslen, Geoffrey. "Chinese Children 'Learn Good Study Habits at Home.'" *South China Morning Post*, September, 20, 2008.

Most Likely to Succeed. Documentary by One Potato Productions, 2015.

Ni Mingjin. "Empowered Administration Program for Rural Schools of Compulsory Education." Posted on the Shanghai Municipal Education Commission website, March 7, 2013.

Tatto, Maria, et al. *Policy, Practice and Readiness to Teach Primary and Secondary Mathematics in 17 Countries.* International Association for the Evaluation of Educational Achievement, 2012.

Tan, Li Hai, et al. "Brain Activation in the Processing of Chinese Characters and Words: A Functional MRI Study." *Human Brain Mapping* 10 (2000), 16–27.

Tobin, Joseph, et al. *Preschool in Three Cultures Revisited: China, Japan, and the United States.* Yale University Press, 1991.

Tucker, Marc. "Why Have American Education Standards Collapsed?" *Education Week*, April 23, 2015.

CHAPTER 13: THE MIDDLE GROUND

China Education Online. "China Becomes the Top Overseas Student Export Country, Total Number Continues to Rise." Posted on ifeng.com, May 7, 2014.

———. "2014 Overseas Studying Trend Report." eol.cn, 2014.

Gao Liang and Li Yaoming. "Seeking the Force for Change: Report on the Education Reform of No. 11 School." *China Education News Network*, April 2, 2014.

Jiefang Daily. "National Day School, Why Do We Choose You?" Posted on xschu.com, February 14, 2014.

Lambert, Phil. "Expanding Horizons: What Can Educators in China and Australia Learn from Each Other?" Paper presented at Nanjing Normal University, China, April 9, 2012.

Lan Fang. "According to a Survey, There Are 100 Thousand People in China Received Non-School Education." Posted on *Caixin Net*, May 14, 2014.

Li Jieren, "Why Can't Other Schools Learn from Beijing National Day School?" March 16, 2015.

Li, Yulan, and Zhenguo Zhu. "Why Does the Rate of College Entrance Test (Gaokao) Participation Keep Declining Year by Year?" *GuangMing Daily*, June 6, 2011.

Ministry of Education of the PRC. "Statistics of Foreign Students Studying in China 2014." March 18, 2015.

Ministry of Education of the PRC, Headquarters of the General Staff, General Political Department. "Guideline for Military Training of High School Students." January 30, 2003.

Nanfang Daily. "Chinese Examination-Oriented Education Started Its Transition from 'Selecting by Points' to 'Selecting by People.'" Posted on news.163.com, July 8, 2015.

People's Daily. "National Day School Breaks the Traditional System of Class Distribution and Changes to Course Selection System." Posted on edu.163.com, February 28, 2014.

Swandon, William, and Brian Kelly, "STEM Proficiency: A Key Driver of Innovation, Economic Growth and National Security." *US News & World Report*, April 23, 2014.

Tencent Education. "2014 Gaokao Admission Scores Released." Posted on qq.com, June 26, 2014.

Wu, Jing. "Number of Students Study Abroad and Returnees Continue to Rise in 2013." Posted on Xinhuanet.com, February 21, 2014.

GENERAL

Akehurst, Jessica Marie. Culture, Cultural Discontinuity and the Need for Change: Understanding Canadian and Chinese Conceptions of Teaching and Learning. Thesis submitted, University of British Columbia, 2012.

Bond, Michael Harris, ed. *Oxford Handbook of Chinese Psychology.* Oxford University Press, 2010.

Carnoy, Martin. "What PISA Can't Teach Us." *Education Week*, February 9, 2016.

Carnoy, Martin, and Richard Rothstein. "What Do International Tests Really Show about U.S. Student Performance?" *Economic Policy Institute*, January 28, 2013.

Chu Zhaosheng, Li Weiwei. "The Ten Key Words of Higher Education for 2015." China Education News Network, December 28, 2015, http://www.jyb.cn/high/gdjyxw/201512/t20151228_647981.html.

Fong, Vanessa. *Paradise Redefined: Transnational Chinese Students and the Quest for Flexible Citizenship in the Developed World.* Stanford University Press, 2011.

Jen, Gish. *The Girl at the Baggage Claim.* Alfred A. Knopf, 2017.

Liang, Xiaoyan, et al. "How Shanghai Does It: Insights and Lessons from the Highest-Ranking Education System in the World." In *Directions in Development.* World Bank Group, 2016.

OECD. *Education in China: A Snapshot.* OECD Publishing 2016.

PISA 2012 Results. *What Makes Schools Successful? Resources, Policies and Practices*, vol. 4. OECD Publishing, 2012.

———. *Creative Problem Solving. Students Skills in Tackling Real Life Problems*, vol. 5. OECD Publishing, 2012.

PISA 2015 Results. *Excellence and Equity in Education, PISA*, vol. 1. OECD Publishing, 2016.

———. *Policies and Practices for Successful Schools, PISA*, vol. 2. OECD Publishing, 2016.

Walters, Kirk, et al. "An Up-Close Look at Student-Centered Math Teaching." *American Institutes for Research*, November 2014.

ABOUT THE AUTHOR

LENORA CHU is a Chinese American author and journalist whose work explores the intersection of culture, policy, and behavior. A former contributing writer for *CNNMoney,* her stories and op-eds have appeared in the *New York Times, Wall Street Journal, Business Insider, Christian Science Monitor,* and on various NPR shows including *Marketplace* and PRI's *The World.* Chu is also an internationally-recognized expert on Chinese education and has given interviews on NPR, CBS, BBC, and CBC. *Little Soldiers,* her first book, won the the American Society of Journalists and Authors' top nonfiction prize, the Nautilus Award recognizing social change, and was a finalist for Stanford's Saroyan International Prize. Raised in Texas, Chu holds degrees from Stanford and Columbia University. She was a management consultant to Fortune 500 companies and has delivered speeches about international education on four continents. For more information, visit www.lenorachu.com.